The Next CIO

How Enterprise Technology Management Powers Autonomous IT

NEXTCIO.BIZ

A. Vincent Vasquez

For additional information, visit nextcio.biz

Cover art by Colin Hoisington
Colinhoisington.com

Copy edits provided by Jessica Ehlert
www.linkedin.com/in/jessica-ehlert

ISBN: 979-8-9854637-3-6

Printed in the United States of America
  Version 1.0 Build 10

Co-Storytellers and Quotes

Alain Brouhard
Former CIO
Coca-Cola Hellenic
Bottling Company

"Companies, corporations and institutions need to further leverage exponential technologies, at the service of the business and for the good of their respective industry or field of play. ETM applications have the power of connecting organizations, even competitive ones, with a purpose of making the world a better and safer place. Vince lays out a vision for how this can be accomplished."

David Ching
Former CIO
Safeway

"Improving operational efficiencies in an IT organization is tough. Vince has laid out an interesting approach around the creation of a new category of application — what he calls ETM — that can help not only CIOs, but also board members to automate and optimize existing manual processes that impact technology within the business more easily and efficiently."

Guillermo Diaz
Former CIO
Cisco

"In order to change the perception of being a technology order taker to a real trusted partner with our own business, we realized the need to transform Cisco IT into a services-based operating model. The ETM process maturity model Vince proposes takes a next step, proposing an approach to help improve the efficiency and security of the IT delivered processes that drive the IT services."

Melissa (Whitney)
Gordon
Current CAO
Tidal Basin Group

"I'm a big proponent of utilizing industry frameworks to help guide technology planning. I especially like how Vince identified the need for CIOs to take a good look at their processes that span technology silos and then offers a maturity framework as guidance on how to improve the efficiencies and efficacy of these processes."

Daniel Hartert
Former CIO
Bayer

"CIOs carry the responsibility to ensure that the 1000's of applications and systems making up an enterprise IT landscape are managed to produce the expected results - on a 24x7 basis, within given performance parameters, defined costs and constant change. But the cold reality is that CIOs and their teams are missing exactly one additional application: a fully integrated toolset that allows to manage their entire landscape, including all related processes and data. Vince provides an interesting response to this dilemma by proposing the building of ETM applications - filling exactly that gap."

Jim Swanson
Current CIO
Johnson & Johnson

"The industry could do a better job of providing CIOs a solution that provides more flexibility to update and optimize our workflows based on changes in the business. And then provide CIOs better visibility into these changes and how they ripple across all our business processes and technology. Vince's proposed ETM approach could be a positive step in that direction."

Mike Kelly
Former CIO
Red Hat

"I think in IT a lot of people in my job have a lot of figurative sheds full of a lot of stuff and they don't know what's in there. I can see where an ETM application like what Vince is proposing could help clean this up. Also, when people talk about this stuff, they used to call it 'technical debt'. Now, I'm starting to hear people call it business debt, because there are all these processes tied to all this stuff that includes all these rules that we put in place that nobody even uses. Having a maturity framework like Vince proposes might be a step in the right direction in helping us optimize these processes — striking the optimal balancing of using automation to remove friction, but still including manual approvals and tasks where it makes sense."

Justin Mennen
Current EVP, Chief Digital and Technology Officer
Rite Aid

"While we've put a great deal of effort into digital transformation, I continue to see many foundational processes built on manual tasks. I think it would be interesting if the industry could develop an ETM application that not only enables automation of these processes, but also provides the visualization of KPIs to quickly notify when a process doesn't execute as planned."

Tim Pietro

Current FinOps and
Technology Business
Management (TBM)
Practice Head
Capgemini Invent

"I like to think TBM is about how do we organize all our assets in IT? How do we take our people, documents, infrastructure – everything we do in IT – and roll it up into a hierarchy to enable us to talk about services and capabilities, such as where are the key areas that we should be investing our IT dollars? The ETM framework Vince proposes is much bigger: it's more about how do we now operate IT?"

Anders Romare

Current CIO and
Senior Vice President
Global IT
Novo Nordisk

"In managing an IT organization with a couple thousand employees, my job is to get all those people driving in the same direction. An ETM application like what Vince is proposing, providing measurements and notifications around our processes, could help me more easily observe the work getting done and monitor guardrails to help me more effectively lead."

Toby Eduardo Redshaw

Former SVP
Enterprise Innovation
Verizon

"The two foundational, often underleveraged, parts of IT are 1) expert, empowered independent program management, and 2) process design and control smartly partnered with both lines of business and finance. Most IT shops are painfully lacking in both areas. An ETM application like what Vince is proposing could go a long way in helping CIOs deploy software to help in both areas, providing automation that is auditable, reliable and unbiased."

Diane Randolph
Former CIO
ULTA Beauty

"Keeping an accurate single source of truth of all the assets used by the business is challenging, especially as manual tasks heighten the risk that errors will creep into our CMDB, and point management tools leave us basically tool surfing to find answers. I like the vision of an ETM application enabling the CIO to manage its entire technology portfolio throughout the lifecycle from one view."

Randall Spratt
Former CIO/CTO
McKesson

"In terms of improving our company's security posture, security frameworks help us know what information we need to know, but the problem for the CIO is how to access this information and manage it on a day-to-day basis. This is very challenging to do using point tools. An ETM application as Vince describes could give CIOs real-time access to this information through automated workflows connected with our existing security management products."

Edward Wustenhoff
Current Vice President
Infrastructure
and Platform Operations
Juniper Networks

"What if the next CIO is not called a CIO? I think this is a very real possibility given the trends in what IT does versus the lines of business, plus the explosion of technology now needed to run the business. I'm intrigued by what Vince is proposing; perhaps it will motivate an industry-wide discussion around not only the creation of an ETM application category, but also the rebranding and repositioning of IT."

Contents

Figures

Foreword

The CIO's dilemma – managing complexity in an ever-changing environment, while also leading and funding disruptive initiatives such as digital transformation.

There is an interesting trend evolving in the world of IT planning and management, where CIOs and CTOs are realizing the implications of an agile infrastructure.

Over the past few years, the role of IT in supporting business has changed dramatically. The focus shifted from traditional on-premises compute to the idea that by being agile, IT could more quickly adapt to business needs. Sure enough this happened, and the implications were profound.

- Cloud workloads are now mainstream, and most (not all) new development is cloud-ready and container-based. Additionally, most enterprises now support multi-cloud environments. Those workloads require new tools, management processes, and policies.
- On premises workloads continue to expand, but often of a different type. These might be mission critical, require high availability or extreme low latency, or are workloads so tightly integrated (or data constrained) that they need to remain in a traditional data center.
- Customer intimacy, especially in retail, is driving IT to move some workloads closer to the customer to reduce that latency and improve the overall customer experience — often to the benefit of brand awareness and corporate reputation.
- Edge is an option for site specific workloads (for example in manufacturing), or a means of pushing specific workloads to remote sites to support a distributed business, or in some cases

to support data location, sovereignty, or regional privacy requirements.

- Work from anywhere — the COVID effect. Enterprises pivoted to a flexible work model during COVID and the implications to endpoint allocation, network access, and application and cloud service provisioning are still being felt.

But the CIOs real dilemma is not harnessing the complexity of this new world or deciding which ITSM, ITAM, ITOM, DCIM or CMBD tool to use, but rather how to take a higher level and more integrated view to assist in leading, operating and securing it all.

For instance, as enterprises move toward hybrid environments, one of the key pain points will center on how to operate IT more efficiently and securely. Organizations have become great at managing silos, but their staff tends to see the world from the construct of endpoints, networks, infrastructure and applications.

And when the organization is managing in silos, it becomes much harder for the CIO to not only gain overall visibility, but also to lead efforts focused on continuous improvement.

In a hybrid environment with a mix of providers, sourcing and architectures, the physical location of an asset (or process) will not be as clearly defined. Yet, its attributes, performance, KPIs and cost will have an increasingly important impact on how IT delivers services to its supported lines of business and end customers.

Ultimately, IT remains responsible for the complex enterprise technology footprint that extends from the home to the data center and out to hundreds of cloud service providers. Along with enterprise technology management processes and organizational changes, infrastructure & operations (I&O) leaders will need tools to actively monitor and manage any process, anywhere, any time.

So, imagine having the ability to visualize a complete purchasing, security or service management process — beginning to end — and to see what components it uses, what performance levels (or issues) they have, what KPIs or SLAs it is hitting, and being able to map all the pieces into a coherent flow.

I had a recent conversation with a CIO who articulated the problem like this:[i]

1. Where is my application? I need to see all my mission critical apps (or even ALL my apps) across all the infrastructure installed in all my data centers, and in the hybrid cloud.

2. What about the enterprise technology management process? I need to see high-level business services and the applications that make them up, plus the infrastructure they are implemented on.

3. What is happening with service performance and availability? I need to be able to visualize system faults, performance metrics, capacity issues, delays, security, and compliance gaps wherever they are.

4. How can my staff figure this out? I need my staff to be able to see the end-to-end connectivity and have the tools to quickly identify bottlenecks or issues that are impacting the end user experience. Why can't I replace staff with automated processes like we have in other parts of the business?

5. And finally, I need to be able to audit and report on everything. Where is the data and when was it last accessed (and by whom)? Which devices are at what levels of software and potentially at risk for zero-day exploits? And which devices, applications, and security level access do each of my

employees have? And how does this map to their current location or job function (since these things will change over time)?

There are a few innovative tools today that are beginning to pull these pieces together, but not many. *The Next CIO* explores the ideas behind Enterprise Technology Management, and how ETM will be a fundamental requirement for IT success in this evolving "infrastructure anywhere" world on a journey toward an autonomous and continuously optimizing IT organization.

David Cappuccio
Former Distinguished VP Analyst, Gartner
August 20, 2022

Preface

I've only known digital.

I remember my first date just after graduating from Carnegie Mellon with a master's degree in computer engineering. It was the early 80s, and I was on a blind date with my father's boss' daughter. I vaguely recall taking her to sushi for dinner. What I do remember was my thoughts driving home after the date: "I was so boring." This thought kept broadcasting in my mind because I had nothing from the analog world to say to her.

For my master's thesis, I wrote a front-end partitioner for a mixed-mode, circuit-logic simulator called SAMSON, essentially a software program to help design digital circuitry. In my first regular job out of graduate school I worked as an engineer for Hewlett Packard. My job essentially was to use technology — built from software programs and test equipment — to fix integrated circuits before going into production. Among other roles, I was the test engineer for the then new 300,000 transistor HP 3000 CPU chip, redesigned from silicon on sapphire to transistor–transistor logic. It would go on to become the first HP PA-RISC processor.

So, on the blind date, I could talk about programming in the "C" language or how excited I was when I used physics to explain that a circuit was failing test because an error in layout was causing latch-up (essentially a short circuit). I remember even my managers went glossy eyed in that meeting. Up to that point, I had been emersed in digital technology, which at the time was a place only visited by nerdy engineers like me.

Needless to say, there was no second date.

Fast forward to today, after working thirty-five years in the tech industry, it's abundantly clear that digital transformation is but one of the latest disruptors on every CIO's mind. In addition, expectations are set: To compete successfully, a company's IT organization needs to deliver outcomes to the business and exceptional experiences to customers and employees, increasingly in a digital format. Consumers groan at the idea of standing in line between 8am and 5pm at a brick-and-mortar anything, such as the DMV to renew a license. We all want the convenience of logging in and doing our business when it's convenient for us.

And our experiences must be quick and painless, otherwise we deride the brand and give it a 1-star rating. Nearly 70% of consumers admitted that page speed impacts their willingness to buy from an online retailer.[ii] Forty-seven percent of consumers expect a webpage to load in two seconds or less, and forty percent will abandon a page that takes more than three seconds to load.[iii]

Quite frankly, I'm often amazed that all this technology works at all, at any speed. When you take a step back and think of the entire stack of technology that must work together to deliver that page load, it's downright impressive. From the software stacks to servers and storage running the applications to electrons flowing across transistors and wires to the network that delivers the data, it's true engineering brilliance that goes into delivering that two-second page load experience.

This is one reason I take it somewhat personally when IT is derided as a tactical cost center to the business — a necessary sieve to the budget and profit and loss.

The rest of the organization would do well to understand that CIOs and their IT departments are really the company's frontline agents for change and innovation for the business. Digital transformation represents just the latest driver of disruptive technologies that the CIO and IT must intelligently adopt for the business to compete.

But it's all very challenging.

We're shooting at a fast-moving target because the technology that CIOs and their IT organizations must manage never stays the same. Sure, with disruptors like digital transformation and cloud migration, some assets previously ran on-premises will shift to SaaS licenses. However, as former Bayer CIO Daniel Hartert puts it: "The shift to licensing technology doesn't make managing the technology easier; rather, it's only going to be more complex going forward. This means how we manage technology also needs to shift."

In fact, most enterprise IT teams today utilize a broad range of technology to enable the business. Each chosen solution solves an important technology management problem. However, because there are many types of technologies and different phases of technology management — not to mention IT organizational silos — IT teams use a combination of management tools: one each for software, endpoints, hardware, on-premises and cloud infrastructure, and so on.

This tool sprawl adds another level of complexity to the technology management challenge, creating data and process fragmentation as the data and processes are driven from different siloed systems.

Historically, increased complexity has given rise to new categories of software applications that simplify processes and drive efficiencies, helping the CIO's peers be more successful in their respective functions:

- CRM provides *lead-to-cash*, enabling sales organizations to manage customer relationships and interactions more efficiently by leveraging large amounts of customer data.
- HCM provides *hire-to-fire*, enabling HR to better manage talent administrative functions.
- ERP provides *procure-to-pay*, enabling finance and other business units to manage day-to-day business activities more efficiently.

In *The Next CIO*, I argue that the modern CIO needs a similar application to simplify the management of enterprise technology (ET) processes that touch a company's broad technology portfolio throughout the technology lifecycle.

- Enterprise technology management (ETM) provides *plan-to-EOL* (end of life), enabling CIOs to define, automate and optimize the processes that derive value from the entire enterprise technology portfolio throughout their usefulness to the business.

I will also argue the need for an enterprise technology (ET) Process Maturity Framework to help CIOs and IT organizations understand where they are on this journey, essentially replacing ad hoc and reactive processes with automated, proactive, and optimized ones — all with an eye on improving the efficiency and security of IT operating processes.

By leveraging an ETM application and the ET Process Maturity Framework, the objective would be for IT organizations to have all their processes that utilize technology defined in software and at least partially automated to fully automated and optimized.

Like the road the automotive industry is traveling, transitioning from operating vehicles with undefined and manual processes to self-driving cars, success would be measured in terms of how close a CIO can get his or her organization to being a fully autonomous IT shop, leveraging an ETM application. Likewise, the vision is a future where CIOs rely on an ETM application to observe, manage and secure enterprise technology processes, autonomously.

That said, a fundamental challenge CIOs face is funding disruptive technology initiatives like digital transformation. But I will argue this budgetary challenge provides just the push the industry needs to create and bring to market ETM applications.

It's time that the "I" in CIO means having fingertip access to accurate information about all enterprise technology and processes that utilize enterprise technology, so IT organizations can more efficiently and securely do their job and be enhanced value creators to the business.

Finally, I'd like to point out that many of the viewpoints shared in this book originate with the co-storytellers. For instance, the above reference to the Chief "Information" Officer came from a discussion with Mike Kelly, former CIO of Red Hat.

My hope is by incorporating these IT leaders' present-day experiences and forward-thinking visions of what type of application could make the CIO's job better, *The Next CIO* stays grounded with actionable insights rather than veering off into science fiction land.

Curious to learn more? Read on and let me know what you think at nextcio.biz.

Vince Vasquez

Acknowledgements

First off, thank you to my "co-storytellers" — the current and former CIOs and VPs who spent valuable time sharing their experiences and insights. Their points of view around the challenges CIOs face and what type of application would help them do their jobs most efficiently and securely provide the very backbone of this book.

For example, Melissa Gordon reminds us of the importance of having a technology plan that provides guidance on the technology portfolio required to meet the needs of the business. It's her point of view that guided me to include the management processes that utilize technology across the entire technology lifecycle, starting with plan to EOL.

She also reminds us how industry frameworks are useful in providing guidelines, best practices, and standardized operating principles to help IT organizations align services delivered with the value they create. For example, if a company stores social security numbers, there are opportunities for breakdowns because the security component isn't just limited to the technology controls that are in place — like encryption — but also includes the people and the processes within the organization. How are the processes that touch personal data handled? How does the business ensure individuals don't have access to data and technology that they don't need access to?

So, it's crucial that IT organizations develop a planned approach to technology spend (and industry frameworks can help here), while also making sure the processes that utilize this technology are more easily managed, governed, audited, and automated. This can help to improve operating efficiencies, reduce costs, and decrease the risk of process breakdowns that can negatively impact the business. It's this point of view that influenced my creating a Process Maturity Framework to provide a standardized way to benchmark the maturity of enterprise technology processes.

Toby Redshaw provided the bulk of the insights that drove the building of Chapter 4: Where's the Money? and Chapter 5: IT is Not Managed Properly. Chapter 4 sets up the tremendous budget opportunity CIOs have if they can find a way to reduce the cost of keeping the lights on. Chapter 5 then organizes what CIOs are tasked to do, shedding light on the fundamental insight that, without an application like Enterprise Technology Management (ETM), CIOs are metaphorically steering IT with one hand tied behind their back. These two chapters provided the set up to my proposed ETM application.

The idea of one company's ETM application being able to connect and collaborate with another company's ETM application along the journey of optimizing enterprise technology processes, lifecycles, and usage, along with helping to keep corporate data more secure, seemed to hit a positive chord with the co-storytellers.

For instance, Alain Brouhard shares the need for even competitors to connect and collaborate across a broad spectrum of topics like security and supply chain, as many organizations face similar challenges. For instance, if my organization can stop a virus before it impacts other companies, including my competitors, that's a good thing. And vice versa. It's a similar story with product quality. If me and my competitors all buy sugar and water from the same supplier that suddenly is having quality issues, it's in all companies' best interest that this issue is flagged and addressed before the quality of all our products is impacted.

Following this line of thought, David Ching is also confident that CIOs would be willing to share experiences and insights with other CIOs. He adds that, to some extent, CIO's already work together participating in different industry and government-sponsored groups. David Ching provides a framework to discuss the role of the CIO (Operate, Innovate, Lead and Secure), which helps to provide an underlying organizational principle for the book.

Justin Mennen adds to this conversation, mentioning that many IT organizations outsource some of their operational workloads to third-party service providers. Automated notifications regarding the status of these managed processes would provide the CIO with broader visibility into the entire IT landscape. Mennen also points out the need for CIOs to have better visualization across their workflows, which influenced the proposed ETM application architecture.

Likewise, Anders Romare — current CIO at Novo Nordisk — pointed out that an ETM application would also need to show KPIs through a dashboard that would enable the CIO to drill down to justify technology investments. He also shared the perspective that in many IT organizations, enterprise technology processes are manual and not measured, so CIOs don't really know how long these processes take to complete. This perspective influenced how the ETM Maturity Levels were articulated.

Jim Swanson pointed out the different layers in the technology stack to consider, such as data, technology, and data science. And all three layers must work together to extract the most value from the technology investments. If one level is in bad shape, or the different levels aren't connected and communicating with each other, then decision making can't be guided by informed data science. His perspective helped guide the need for connectors between the ETM application and point management tools so IT organizations wouldn't have to stitch together custom solutions to maximize value from technology spend.

Daniel Hartert reminds us that even though there is a shift from running technology on-premises to licensing from third-party vendors running in the cloud, it doesn't make managing the technology any easier for the CIO. For instance, even though the CIO may no longer be responsible for change management of SaaS offerings, the CIO is still responsible for ensuring that the performance, availability, and security of these offerings continue to meet the needs of the business, all without having underlying control of the technology components. Likewise, the CIO is still responsible for the processes that ultimately utilize these technologies, regardless of whether they're running on-premises or in the cloud. Hartert points out this reality could be a motivator for why the industry should consider developing an ETM application offering to become "the CIO's application."

Diane Randolph shared candid examples of how challenging it is for her organization to maintain an accurate inventory of all their assets in their enterprise technology portfolio, including devices deployed in over 1,200 stores. Diane shares there were point tools offered by the hardware vendors that could be used to help manage the various pieces of technology, but what she really wanted was all their technology portfolio to be managed throughout the lifecycle from one view. This approach would remove complexity and improve the process efficiency for operations. This is precisely what an ETM application would provide.

Mike Kelly approaches the discussion around automating enterprise technology processes with a unique perspective. As the former CIO of

Red Hat, which offers the automation tool Ansible, Kelly's organization also uses the tool in-house to help automate internal processes. He reminds us that even when applying automation, we can't lose sight of the critical role employees continue to play in enabling the business. Kelly shares that one of his objectives is to help employees be EPIC (i.e., effective, productive, innovative, and collaborative) in every function, increasing operational excellence within the company. Among other areas, this point of view provided motivation and backdrop for the discussion around how an ETM application could also be utilized to improve employee experiences.

Guillermo Diaz shared his experiences transforming Cisco IT from a client-funded operating model to a services-based one, delivering IT-as-a-Service. He helped make the connection between an IT service, such as a quoting service, with the IT operating process that delivers the end-to-end solution, such as *quote-to-cash*. This connection is critical because it drives the point that the CIO not only has to guide the organization to improve efficiencies in delivering IT services, but they would do well to also keep a focus on improving process efficiencies.

Speaking of delivering IT services, Tim Pietro helped articulate the complementary nature of Technology Business Management (TBM), which among other things provides a framework and methodology focused on cost transparency, identifying the total cost of IT, and shaping demand for IT capabilities with an ETM application and framework that focuses squarely on how to improve how the CIO operates IT.

Edward Wustenhoff helped to take the discussion all the way to wondering if the next CIO will even be called CIO, or will the industry decide to rebrand the name CIO and IT organization to something different. Afterall, "information technology" was first coined way back in a 1958 Harvard Business Review when authors Harold J. Leavitt and Thomas C. Whisler said "The new technology does not yet have a single established name. We shall call it Information Technology." Considering that more than six decades later we've seen an immense change in the use of technology in the business, perhaps it is time IT received some rebranding.

I'd also like to thank Dr. Timothy Chou for encouraging me to write this book. As a former president of Oracle Corporation, more than 25 years as a lecturer at Stanford University and current board member at Blackbaud and Teradata, Dr. Chou is well versed in the life of IT. He provided guidance throughout this process, taking the time to listen to my ideas and on several occasions helped steer the book's direction with his invaluable insights.

Finally, I'd like to thank my son Diego for listening to me ramble on about writing this book during our almost daily walks together. As a bright fourteen-year-old, he made up for his lack of understanding of the CIO's world by listening intently about his daddy's work. And on more than a few occasions, offered some very meaningful advice.

Chapter 1: Funding Digital Transformation

What About the "What Abouts"?

As a CIO, your team is no doubt constantly barraged with the "what abouts."

What about Web 3.0? What about cryptocurrency? What about security? Why aren't we doing more in artificial intelligence and machine learning? What about IOT? What is our cloud migration strategy?

And what about digital transformation — something that every CIO has firmly on their must-do list?

You know the Blockbuster versus Netflix story backwards and forwards. You've heard people say you're operating more like Blockbuster and need to be more like Netflix. Online. On-demand. It's all about delivering exceptional digital experiences for customers and employees.

The CIO's dilemma: Where's the budget?

But it takes money and headcount to digitally transform your business. It takes money to investigate and act on all the "what abouts" potentially relevant to helping your IT organization deliver more value to the business.

So, where's the budget going to come from?

The CFO certainly isn't handing out money like Halloween candy. And you've already done several cost-cutting rounds. You no doubt have tried to squeeze as much cost from your vendors as possible. And you've likely outsourced at least some of your organization's workload to IT managed service providers that can do your "mess for less."

In *The Next CIO*, I will show you that you don't need more budget from the CFO to fund transformational initiatives. There's still plenty of blood in the stone.

In other words, *The Next CIO* is first and foremost about cost management. How can a next-generation CIO take more cost out of the IT organization's budget to fund "what abouts" like digital transformation? That said, this book is also about improving IT efficiencies and data security that are, in a way, byproducts of running a smarter IT organization that can reduce operating costs. CIOs who run a smarter IT organization also gain improved observability into the processes used by the team and supported business entities to run the parts of the business that leverage technology. This enables the next CIO to better articulate the IT organization's value creation back to the business.

But first, back to cost management.

The disruptor imperative

So, why is finding budget to fund the "what abouts" important?

Simply put, any company tasked with digital transformation that is funding its business through its own operating income is at tremendous risk. This is essentially any incumbent organization.

Stories of failed companies that did not address similar risk are all too familiar for anyone who has lived in the tech industry: A company goes from market leader to virtual non-existence, haven fallen prey to a disruptor that redefined its category. The Blockbuster to Netflix story is just one of many.

In 1999, I attended a meeting at a Nokia R&D building in Tampere, Finland to discuss next-generation mobile phone technology. As the worldwide leader in mobile phones at the time — with a heritage of bringing leading phone technologies to market — Nokia was confident that adding cameras to their mobile phones would disrupt camera industry leaders like Canon. Fast forward to 2007 when Steve Jobs launched the iPhone designed first as a computer and second as a phone, disrupting the smartphone category. History speaks for itself.

We all know that digital transformation is another market disruptor, as digital technology innovation destabilizes legacy business environments and models. Smartphones and video streaming are only two familiar examples of how disruptive technology made legacy businesses also-rans.

Incumbent organizations with pre-existing business models are at significant risk of value loss if they don't digitally transform themselves to better compete in the digital era against born digital competitors. These days, it's relatively easy for a start-up to secure funding off a PowerPoint presentation, scale infrastructure running a credit card at a public cloud provider like Amazon Web Services or Microsoft Azure, scale labor through freelancing platforms like Upwork, and scale manufacturing through contract manufacturers like Alibaba.

Similarly, with the acceleration of competition, incumbent companies also must worry about their incumbent competitors that more quickly and successfully digitally transform themselves.

The good news is that to varying degrees, all CIOs have known for at least a few years now that they need to digitally transform their organizations. No surprises here.

The bad news lies in the budget: How do CIOs from incumbent companies filled with legacy operating models and processes fund their digital transformation initiatives?

Zone to Win Guideposts

To not only survive but thrive through digital disruption, an organization must rethink its entire operating model and adapt, incorporating new technologies and infusing digital innovation into the entire supply chain.

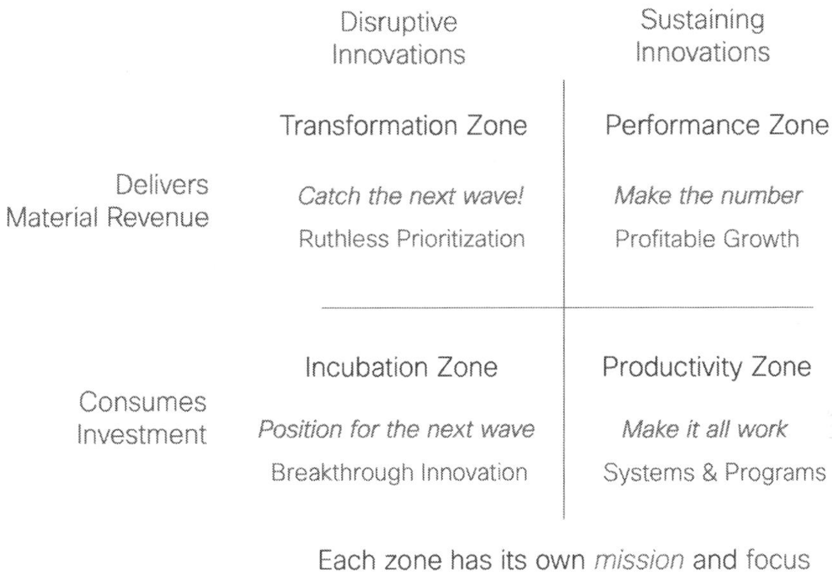

	Disruptive Innovations	Sustaining Innovations
Delivers Material Revenue	**Transformation Zone** *Catch the next wave!* Ruthless Prioritization	**Performance Zone** *Make the number* Profitable Growth
Consumes Investment	**Incubation Zone** *Position for the next wave* Breakthrough Innovation	**Productivity Zone** *Make it all work* Systems & Programs

Each zone has its own *mission* and focus

Figure 1-1: Zone to Win Management Framework

Fortunately, for those executives who have embraced the need to digitally transform but are still struggling to find the means, Geoffrey Moore, in his book Zone to Win[4], provides a helpful framework. As shown in Figure 1-1, this framework identifies four zones within incumbent organizations that require different management models.

The Performance Zone

The Performance Zone represents the revenue-generation part of the business, tasked with making the quarterly numbers and meeting investor expectations. This is where customers are served and provides the operating income that funds current and future operations.

The Productivity Zone

Operations in the Productivity Zone keep the lights on and make the Performance Zone work efficiently and effectively, as well as ensure the company complies with all relevant governing compliance regulations. Essentially, everything you don't charge a customer lives in the Productivity Zone, such as HR, finance, IT, security, and marketing. It includes all the cost center functions that are subtracted from the Performance Zone revenue minus the cost of goods sold to produce the company's operating income.

In a typical year, perhaps 90% of a company's budget is spent in the Performance and Productivity Zones.

The Incubation Zone

The Incubation Zone is where skunk works projects live, as companies test out potentially disruptive technologies that have the potential to impact the business sometime in the future. In the old days this Zone was often called a "Lab", such as Sun Labs where Java was invented. Today, companies generally leverage a venture start-up incubator for this type of work.

This work is done outside of the Performance and Productivity Zones, as the technology is not ready to scale, so it shouldn't be incumbered by all the requirements of a mature offering. Also, the impact to the overall budget is typically minimal, so we'll ignore Incubation Zone funding for the purposes of this discussion.

As a side note, a business that has no disruptions in its segment can operate within only these three zones. However, with digital transformation impacting practically every sector, along with other driving disruptors such as cloud migration, IOT and hybrid workplaces, this is rarely the case.

This leads us to the Transformation Zone.

The Transformation Zone

The Transformation Zone is where disruption occurs, either because the company is disrupting the market or a company's business is being disrupted. The Transformation Zone is where enterprise-wide change occurs at scale, creating a new source of revenue for the business. This is where digital transformation initiatives live.

If your company is being disrupted, then only paying lip service to needed Transformation Zone initiatives will likely result in losing customers to the disrupting companies, either because they offer better digital experiences or because they offer better pricing as they run more efficient, digitally enabled operating models, such as by leveraging automation and AI/ML-driven insights.

Three Options to Fund Digital Transformation

Raid
operating income

Squeeze
operating costs

Improve
ET process efficiencies

Figure 1-2: Three options to fund digital transformation

The big challenge here is how to fund the transformational initiatives, like digital transformation and the rest of the "what abouts." Budget must come from somewhere.

Raid operating income

For incumbent companies, one place to find funding for staff and to accelerate transformational efforts is by dipping into operating income. After all, the money is sitting right there in the bank if the company is profitable. However, this approach results in lower recorded profits that can drive investors right into the arms of the disrupting competitors.

So, funding through operating income is an option, but not a great one.

Squeeze operating costs

Another place for the CIO to find budget to fund transformational initiatives is on the cost side of the balance sheet, especially as it relates to activities in the Productivity Zone.

For instance, a company can use an IT-managed service provider to outsource its existing operating workload like running the email system, managing the network, or re-architecting the cloud-destined legacy applications. A company can also outsource emerging activities like security so additional headcount doesn't have to be hired.

This can be a good option, as managed service providers gain cost advantages through specialization and the use of manual labor sourced from lower wage geographies. By taking this route, the CIO can repurpose operating headcount to projects that are seen as more strategic for the company. And then every three years or so, the CIO can issue new RFPs, continually trying to lower the fees charged by vendors for their outsourced workloads.

Anders Romare, current CIO and Senior Vice President of Global IT at Novo Nordisk, shares how it can be quite effective putting a new RFP out to bid, and allow competition amongst vendors to bring costs down.

However, most IT organizations have already completed multiple rounds of these types of cost cutting measures, so don't expect to find much low hanging fruit here. Certainly, there may not be enough potential savings available to free up significant budget to accelerate success exploring and adopting "what about" disrupters, that is unless something truly transformative can be applied to radically improve IT productivity.

This takes us to the third option.

Improve Enterprise Technology Process efficiencies

The next way to look for cost reduction is by leveraging digital transformation itself to transform the CIO and IT organization's processes and associated workflows. This option enables the CIO to improve IT productivity by digitally transforming enterprise technology (ET) management processes — those process that touch the entire technology portfolio deployed by the overall company.

Further, digitally transforming enterprise technology processes and workflows will also improve the quality of service that the CIO's IT organization delivers back to the business, indirectly improving

Performance Zone-generated revenue. For instance, customer experiences for companies with retail outlets could be improved if the ET processes more efficiently ensured that point-of-sale (POS) devices are always working and up to date, so the maximum number of checkout lines were always available.

Let's explore this third option in more detail, starting by looking at the maturity of existing enterprise technology processes.

Figure 1-3: ET = Enterprise Technology (not extraterrestrial)

Chapter Summary

Key takeaways
- It's imperative that CIOs further reduce operating costs to free up budget to fund disruptive initiatives such as digital transformation.

The good news? There is more cost available to squeeze out of a CIO's existing IT budget.

- Zone to Win by Geoffrey Moore provides a helpful framework to organize thinking around IT costs, placing IT initiatives into four categories: Productivity Zone, Performance Zone, Incubation Zone, and Transformation Zone
- The next place for the CIO to find cost reduction within the IT organization's budget is at the enterprise technology process and management level.

Key questions

- What is the list of what abouts technology disruptors your IT organization should investigate?
- How much budget and headcount have you assigned to each one of these what about disruptors?
- For each what about technology disruptor, is the provided budget below, at, or above required levels and by how much?

Chapter 2: Rating ET Process Maturity

For most incumbent companies of reasonable size, the ET process landscape is inherently complex, as technology impacts nearly every operational process from purchasing and keeping data and employees secure to maximizing the value attained from the technology purchases.

Take the routine process of secure offboarding an employee as an example — from employee termination set to all access revoked and company-issued technology returned. This *separation-to-recovery* process is described with the following steps:

- **Separate**: Terminate employment in HR system such as Workday; notify IT, HR, finance, legal and other functional stakeholders of the change in employee status; handle for voluntary or involuntary separation
- **Deprovision**: Lock corporate laptops and mobile devices; revoke access to endpoints, network, infrastructure (cloud) and SaaS/on-premises applications such as Salesforce, Microsoft 365 and SAP; cover access within and outside the purview of single sign-on (SSO); preserve employee's previously accessed data and workspaces
- **Reassign**: Transfer ownership of and access to documents, data, cloud resources and other work product to others; setup mail forwarding and auto-replies; delete recurring calendar invites

- **Recover**: Initiate returns of corporate devices and accessories; reclaim cloud resources and return application licenses to entitlements pool; enforce legal hold and data preservation requirements
- **Reallocate**: As appropriate, reimage and return employee issued devices back into inventory
- **End of Life**: As appropriate, sanitize and wipe devices; destroy, donate or recycle; store certification of destruction; update financial audit manual and financial systems for audit readiness

Using the eye test, a complete employee offboarding process is complicated, touching many organizations and technologies. Partially because of this complexity, this process is often implemented through multiple workflows that contain many manual tasks — and require different personnel to use different, siloed technology management tools. This approach is inherently more expensive and more prone to human error than if the tasks were automated across the IT management tools.

From technology plan-to-EOL

In addition, ET processes need to manage an organization's technology portfolio throughout the entire technology lifecycle, from technology planning to procurement through EOL. For instance, an accurate budget requires accurate demand forecasting. The required input is not only planned new technology purchases, but also procurement of replacements for broken technology and upgrades to technology reaching end of life.

Tidal Basin Group CIO, Melissa Gordon, reminds us: "It's important not to miss technology planning as a guide to IT-related decision making. This is where regulatory bodies and frameworks come into play. For instance, depending on what a company does, it might leverage several security frameworks to translate its requirements into a technology infrastructure that is more automated and, therefore, more easily audited. Right now, a lot of this technology planning is highly manual, which creates many opportunities for breakdowns because the security component isn't just limited to the technology controls that are in place but also to the people and the processes within the organization."

If the CIO ensures a technology plan is developed and periodically reviewed, technology spend can be better aligned with all technology-leveraging processes run by the business. The technology plan and processes then drive KPI measurements utilized by all levels, from the board to the process owners.

Foundational to developing a technology plan is knowing what technology is already in place. It's like when I needed to put together a plan to test a production run of integrated circuits, but no one could tell me how many testers would be available. Frustrating.

However, for many companies, it's difficult enough to keep an accurate inventory of the entire technology portfolio, let alone accurately forecast and plan for what technology will need to be replaced over the next few quarters. Surprise technology purchases can wreak havoc on fixed budgets.

So how does the modern CIO and broader IT organization measure how efficiently they are planning, managing, controlling, and optimizing technology and resource spend, while also enabling the business to continue delivering exceptional experiences and outcomes that are dependent on this technology?

The ETM Framework

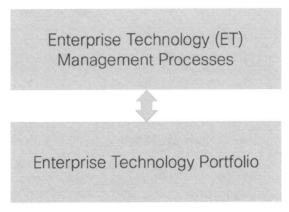

Figure 2-1: ETM Framework

To help us answer this question, we'll leverage the Enterprise Technology Management (ETM) Framework, as shown in Figure 2-1. This simple but powerful framework helps us focus our attention on the fact that there are processes the company uses to run the business that touch the broad inventory of technology:

- **Enterprise Technology Management Processes** are the processes used to run the business that touch the company's enterprise technology portfolio.
- **Enterprise Technology Portfolio** is the entire inventory of technology used by the business.

Let's look closer at each of these components, starting with the technology.

Enterprise Technology Portfolio

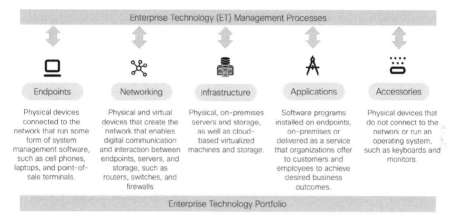

Figure 2-2: Enterprise Technology Portfolio

It goes without saying that technology is a critical resource in today's enterprise. But the sheer number of different technologies that businesses rely on make the job of efficiently and securely managing these resources throughout their usefulness to the organization quite complex.

To help with this complexity, the ETM Framework segments all technology into five broad categories, as shown in Figure 2-2.

1. **Endpoints**: Physical devices connected to the network that run some form of system management software, such as mobile phones, laptops, and point-of-sale terminals.
2. **Networking**: Physical and virtual devices that create the network to enable digital communication and interaction between endpoints, servers and storage, such as routers, switches, and firewalls.

3. **Infrastructure**: Physical, on-premises servers and storage, as well as cloud-based virtualized machines and storage.

4. **Applications**: Software programs installed on endpoints, on-premises or delivered as a service that organizations offer to customers and employees to achieve desired business outcomes.

5. **Accessories**: Physical devices that do not connect to the network or run an operating system, such as keyboards and monitors.

Enterprise technology management processes are then applied to maximize the usefulness of this technology to the business.

Enterprise Technology Management Processes

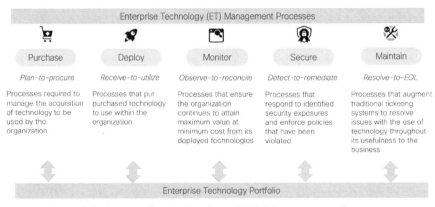

Figure 2-3: Enterprise Technology (ET) Management Processes

Enterprise Technology Management Processes include all the processes managed by the IT organization that utilize technology to meet the needs of the business. At a high level, the objective of these processes is to deliver maximum value to the organization through optimal use of technology, budget, and associated headcount resources required to run the processes.

As shown in Figure 2-3, the ETM Framework segments these processes into five general categories that track the technology lifecycle.

1. **Purchase Management** (*plan-to-procure*): The processes required to manage the acquisition of technology to be used by the organization.
2. **Deploy Management** (*receive-to-utilize*): The processes that put purchased technology to use within the organization.
3. **Monitor Management** (*observe-to-reconcile*): The processes that ensure the organization continues to attain maximum value at minimum cost from its deployed technology.
4. **Secure Management** (*detect-to-remediate*): The processes that act on identified security exposures and enforce policies that have been violated.
5. **Maintain Management** (*resolve-to-EOL*): The processes that resolve identified cases involving issues with the use of technology.

As a side note, these ET processes will most likely operate in the Productivity Zone. Even though they will touch technology deployed in support of the business operating in all four zones, the actual processes themselves will be more aligned to workflows that help keep the lights on and manage technology used by all facets of the business.

ET Process Maturity

The maturity of the ET processes deployed by the IT organization will say a lot about how efficiently and securely the business can run those processes.

At the lowest level of maturity, a business has no process defined and the work is being done ad hoc, reacting to issues and likely following steps stored in an employee's head — an employee who someday may leave for greener pastures.

On the other end of the maturity spectrum, a process is not only fully automated but also continuously optimized to run as efficiently as possible. A process at this level of maturity is likely automated and proactively predicting situations so actions can be taken before an issue negatively impacts business outcomes or user experiences.

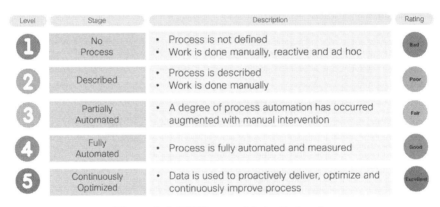

Figure 2-4: ET Process Maturity levels

An ET process can run at one of five levels of maturity, as shown in Figure 2-4:

- **Level 1: No Process** – The process is undefined. Work is done completely manually, is ad hoc and reactive. This is the most inefficient state, so Level 1 gets a bad rating.
- **Level 2: Described** – The process is described but implemented manually. Manual processes are inherently inefficient, so Level 2 gets a poor rating, as completely manual processes in general are

more expensive and take more time than letting computers do the work through automation.

- **Level 3: Partially Automated** – The process utilizes automation to make it more efficient, but still depends on manual intervention to complete. Because some level of automation is involved, Level 3 gets a fair rating.
- **Level 4: Fully Automated** – The process is completely automated and completion time is measured. Given the process requires no inefficient manual intervention, Level 4 receives a good rating.
- **Level 5: Continuously Optimized** – The process uses data to optimize and continuously improve its activities. It is executed, monitored, and managed through dynamic workflows, making decisions based on output from automated tasks, adapting to changing circumstances and conditions, and simultaneously coordinating multiple tasks. Level 5 receives an excellent rating, as processes at this level are running as efficiently and securely as possible.

Of course, not all ET processes will be candidates to be fully automated. For instance, some processes will require manual approvals, a physical activity to occur like returning a company issued laptop or might be too complex to fully automate.

In addition, at any given time, an IT organization's various processes will realistically be running at different maturity levels.

However, the objective should be to describe, automate and continuously optimize all the processes IT is responsible for en route to running an autonomous IT shop. This is where, as much as possible, the CIO can reduce the cost of keeping the lights on.

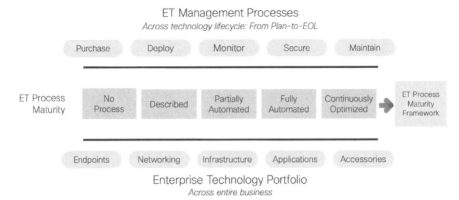

Figure 2-5: ETM Framework with ET Process Maturity levels

Figure 2-5 shows how these five levels of ET Process Maturity fit into the ETM Framework.

Of course, a framework is only useful if it can inspire and motivate change that brings more success to the organization. This is where the ET Process Maturity Framework will come into play. Driven by the ET Process Maturity, the framework will provide a way for the CIO to identify specific opportunities to improve ET processes to ultimately deliver more value back to the business and help fund disruptive initiatives like digital transformation. This will be covered in *Chapter 7: ET Process Maturity Framework*.

But first, let's discuss a big question: Why aren't all ET processes operating at the highest level of possible maturity to begin with?

Barriers to ET Process Maturity

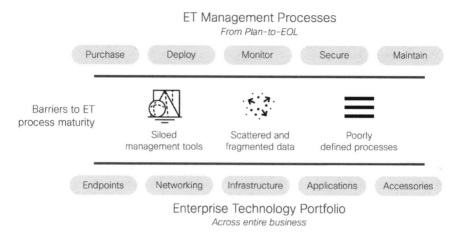

Figure 2-6: Three barriers to ET Process Maturity

As shown in Figure 2-6, IT organizations face at least three barriers to ET processes operating at maximum maturity: siloed management tools, scattered and fragmented data, and poorly defined ET processes.

Siloed management tools

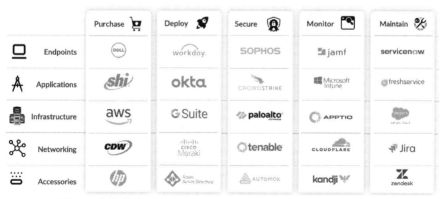

Figure 2-7: Point tools used to manage technology portfolio

For the most part, all people in the CIO's IT organization work in their own silos, taking care of the work needed in each domain of the organization. Unsurprisingly, each siloed group adopts its own set of tools to help bring efficiencies to their jobs, with examples shown in Figure 2-7.

Because there are many types of technologies and different phases of technology management, IT teams use a combination of management tools: one each for software, endpoints, networking, on-premises and cloud infrastructure, and so on.

For instance, the folks responsible for the network will no doubt use a network management tool to help them manage all the complexity involved in keeping a network running. The DevOps folks will likely use application management tools to help them ensure optimal user experiences.

The CISO's team will no doubt have its own set of security tools to mitigate vulnerabilities, enforce policies, and triage issues. In fact, according to one estimate, the average enterprise deploys over 130 security tools, while a medium-size company relies on 50–60 tools on average to help keep the organization secured.[5]

Many of these siloed groups implement some level of task automation to help bring efficiency to their individual jobs. This is a good start but it is not the same as automating entire ET processes.

If an IT organization wants to automate a process that touches various technologies, one option is to stitch solutions together themselves by doing their own custom integrations. This approach is not only time consuming (of IT resources that are already in short supply), but it also risks incurring additional costs as the custom automation needs to be supported and maintained as the point tools update.

Another option is to follow various vendors such as SAP as they expand their process footprint beyond being considered point products. For instance, SAP offers several processes that expand beyond traditional ERP, such as recruit-to-retain Human Resource Planning (e.g., Talent Acquisition & Onboarding, Personnel Administration and Performance & Succession), Financial Management (e.g., Field Sales Monitoring, Customer Profitability and Vendor Procurement Monitoring), and Customer Asset Management (e.g., Equipment Maintenance, Equipment Refurbishment and Equipment Disposal). That's quite a broad range of ET process functionality.

However, it seems like a very tall ask for SAP to operate as a best-in-class offering both across all ET management processes (including security) and the entire enterprise portfolio from endpoints to infrastructure and applications. For instance, to expand into human resources, SAP bought SuccessFactors. However, as Alain Brouhard — former CIO at Coca-Cola — shares, a company might prefer to use a different HRM application and not be forced into an SAP-dominated process world.

Plus, it's difficult to imagine SAP — or any other company — buying point products across the entire ETM framework, many of which likely wouldn't be for sale, such as cloud provider management software including AWS Systems Manager.

Scattered and fragmented data

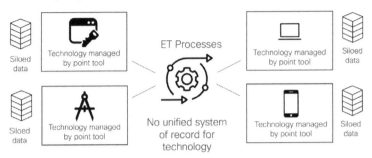

Figure 2-8: Data about technology is scattered and fragmented

One consequence of using point tools is that an enterprise's data about technology gets scattered across the siloed tools with no centralized system of record. Since point-tool functionality varies, they often have limited data that only concerns the technology function they are managing. This results in fragmented and often inaccurate enterprise technology information, making it difficult for a business to understand the state of its entire technology portfolio.

For instance, the CIO of a healthcare provider had to manage the technology for hundreds of dental offices across the country. As part of this service, the CIO's team manages each dental office's networked x-ray machines (endpoints) throughout their lifecycle. If this process were automated, the IT team might be able to:

- Track which systems are approaching end of life, so they can be scheduled to be replaced before offices struggle with outdated equipment
- Forecast and budget for replacement x-ray equipment, avoiding any budget request or supply chain surprises
- Ensure newly purchased systems are configured correctly and patches are maintained

- See and respond to detected security threats and policy violations
- Track the historical reliability of each of its deployed systems, which can be useful in working with their x-ray vendor

Today, the unfortunate reality is that most IT organization's processes are not automated or optimized. For instance, most IT organizations can't click a button and receive — within a few minutes +/- a second or two — an entire inventory of all technology deployed by the organization with exact locations, owners, lifecycle stage and security state. This would be handy for that healthcare provider to help them manage those networked dental office x-ray machines, and to ensure the operational availability for the dental practices they serve along with exceptional experiences for their customers.

Instead, it's typical for different IT staff members to log in and out of several point solutions to access the data required to run various ET processes. This requires several manual interventions, including manually normalizing, aggregating, and analyzing disparate data sets. In the case of our healthcare provider, the risk is that it may take a long time to remediate critical performance or operational issues on some x-ray machines, while new x-ray machines may not be ordered on time to meet the needs of their customers, thus fueling ET process inefficiencies.

Poorly defined ET processes

Speaking of processes, the unfortunate truth is that many IT organizations have not defined all their ET processes. Or if they have, they tend to be defined in lengthy PDF files that are hard to define and follow and even more difficult to govern, maintain and share consistently.

As Daniel Hartert shares: "The reality is that due to many operating pressures, people often didn't have the time to follow processes exactly. Plus, there are too many exceptions because the world is more complex than considered in the written-down processes. So, when it's announced that auditors are coming in six weeks, everybody starts working like crazy to update log files, protocols and so forth to prove that we are living the system."

This frenzy to pass an audit just reduces the operating efficiency of the IT organization.

For maximum IT efficiency, processes would ideally be defined in software. This way, not only can they be automated, but details around what tasks were completed and when can be precisely tracked.

Lack of ET Process Maturity means opportunity

On the other hand, the fact that existing ET processes are not described or lack relative maturity implies there is room for improved efficiencies, and therefore, great opportunity for the CIO. If enterprise technology processes can be made more efficient through automation and continuous optimization, more money can be saved from existing IT budgets, which could be reallocated to fund Transformation and Performance Zone initiatives.

The next question to ask is, why is having immature ET processes such a bad thing? Is it just all about wasting a few dollars, or is there more at play here?

Chapter Summary

Key takeaways

- The ETM Framework provides a starting point to help understand the maturity of each process that touches the technology portfolio.
- Each ET process is on its own maturity journey from undefined to automated and optimized.
- CIOs would do well to recognize the barriers — such as siloed management tools, scattered and fragmented data and poorly defined processes — that impact ET Process Maturity.
- Unless an organization's ET processes are all automated and optimized, there is room to improve their efficiencies and find savings that can be applied to Transformation and Performance Zone initiatives.

Key questions

- What is the full inventory of your organization's technology portfolio?
- Which ET processes touch this technology portfolio?
- What is the maturity rating for each of these processes?

Chapter 3: Immature ET Processes are Bad

To the optimist, the glass is half full.
To the pessimist, the glass is half empty.
To IT, the glass is twice as big as it needs to be.

For many working outside of IT, this is unfortunately the perspective of how the IT organization operates.

Equally unfortunate is that at least some of this perspective may be well founded, as a typical IT organization inefficiently manages its processes that touch its complex technology landscape through a broad collection of siloed tools, scattered and fragmented data, and manual intervention.

Among other things, this approach is very human-labor intensive, time-consuming, and error-prone. This decreases IT productivity, adding to the costs of keeping the lights on in the Productivity Zone, and missing the opportunity to leverage data to make more informed decisions.

In addition, if the processes aren't well-defined and followed, this could lead to problems during audits. Perhaps even more unfortunate is that inefficient, manual processes can make it difficult for the CIO to articulate IT's value creation to the company.

Let's explore these points in a bit more detail, as they represent some of the specific problems each CIO needs to solve if money will be found in the existing budget to fund what about disruptor initiatives like digital transformation.

| Manual means expensive, slow, tedious, error-prone and depleted workforce experiences | Poorly-defined processes create audit, compliance and security risks | Challenging to articulate IT value creation to the business |

Figure 3-1: Why are immature ET processes so bad?

Manual Processes are Expensive in More Ways Than One

It goes without saying that deploying manual labor to do IT tasks is more expensive than using automated processes run by computers. Unfortunately, many IT workflows are still manual.

According to one survey[6], 52% of organizations handle repetitive ITSM tasks like user management, employee onboarding / offboarding and employee job changes manually. Another 38% have implemented automation, but only through a patchwork of custom-built scripts and APIs. Just 10% of organizations say they leverage an integration and workflow platform to handle these tasks automatically.

That said, the incremental costs of using manual processes go beyond simply using expensive IT headcount to do tasks that computers could do more cheaply.

Overinvesting in IT infrastructure

For instance, let's look at a more specific example such as the process of internal technology demand forecasting within procurement. In this use case, requests for new technology purchases generally come in manually from different departments, expressed by the multitude of areas managed by siloed groups within IT.

Folks who manage the network request more routers, firewalls, and access points. The CISO office asks for new security tools. Folks managing infrastructure might come in saying they need to purchase more services for their virtual environments.

This process is generally not integrated at all within most companies, leading to higher costs due to higher levels of IT investments than needed.

As Daniel Hartert shares: "This is because nobody in the various IT departments wants to be in a situation where daily operations capacity requirements exceed what they have installed. In a way, IT departments are motivated to install more assets than what is needed, and sometimes much more. When you think about real-time supply chain, commonplace for a modern manufacturer such as in the automotive industry, a Mercedes Benz factory would never be able to afford to put so many components in stock just in case they might need them in the future."

Unoptimized endpoint lifecycle management

Diane Randolph, former CIO of leading retailer Ulta Beauty, shares some of her technology management challenges. Diane and her IT organization were responsible for managing the technology used in the corporate office, six distribution centers and in over thirteen hundred retail stores. Each retail store deployed many types of technology such as point-of-sale registers, cameras and traffic counting sensors.

With the corporate office and distribution centers, someone from the IT organization was physically placed in each location to help with managing the technology. However, it was not financially feasible to place an IT or field service person in each retail store.

Diane recalls when she and her team did a large-scale rollout of mobile technology to all Ulta retail stores at once. The retail stores were thrilled with how fast the endpoint devices were delivered and put into service.

But then came the challenge of managing the lifecycle of these devices. Diane shares: "In previous days when we were mostly managing the lifecycle of servers and corporate laptops, this was built into our processes and relatively easy to do. But when you're talking about thousands of relatively expensive mobile devices with relatively short life expectancies, this moves the problem of lifecycle management to a new level of complexity."

Ulta deployed a mobile device management (MDM) and CMDB to provide the single source of truth for all the mobile devices. So when a device was powered off and no longer connected to the MDM, Diane's team understood this could mean there was an issue with the device in the store that had not yet been reported, the device simply needed to be charged, or it had been stolen (although this was not a problem Ulta experienced).

Regardless, the process to figure out what was going on was highly manual and inefficient, as someone from the IT organization had to essentially call and ask a store associate to spend time locating the device instead of serving customers. Given the devices were networked, if Diane's team could simply look up the last known data on the offline device, like the status of its battery and its last known location, this problem could be alleviated. Likewise, from a security perspective it would have been helpful to have an automated workflow trigger if the device went offline, such as locking it until it was cleared to be in known hands.

However, more challenging was modeling the refresh of these devices through their lifecycle. Given all the devices had been purchased at the same time in support of the rollout, Diane knew this created a liability because all devices would reach their end of life at approximately the same time. This could create a big hit on the budget, taking money away from other, more innovative initiatives.

What seemed to make the most sense was to operate a rolling upgrade to newer hardware versions, insulating the company's budget from a massive replacement of a large population of assets all at the same time. Diane would have liked to have an automated process running that intelligently and cost-effectively ran this type of workflow. For instance, ideally this process could model different potential trade-offs. For instance, what if they refreshed the mobile devices but delayed server refreshes for six months? Or vice versa?

Plus, with so many devices deployed, it was inevitable that some percentage would fail before their scheduled EOL, so this data also had to be considered when modeling the repurchase of replacement devices. Diane would have liked to have a tool that leveraged knowledge of the technology reliability to provide insights to help her make more informed decisions regarding the optimized mix of age of equipment, while also predicting failure rates over the next twelve months. With these insights, she could purchase a certain number of devices in anticipation of field failures, preventing costly emergency purchases.

Diane shares that her team did leverage the point tools offered by the various hardware vendors to help with the management of the various pieces of technology. But this was all done in silos, making it impossible to automate the modeling of optimized purchases that spanned across the entire technology portfolio. What Diane really wanted was to have all of Ulta's technology portfolio managed throughout the lifecycle from one view. This approach would remove complexity and improve the process efficiency for operations.

Shadow IT makes keeping technology inventory and budget difficult

Since the introduction of cloud computing — which introduced the practice of running your credit card to purchase infrastructure — the CIO has had to contend with the additional complexity of managing decentralized purchases of IT services and assets.

This shadow IT makes it harder for the CIO to roll up a predictable IT budget. However, when employees who ran their own company-issued credit cards leave, how does IT efficiently make sure that the monthly cloud infrastructure charges or SaaS licenses are either turned off or properly transferred to another employee?

If SaaS software or cloud resource discovery, ownership, security, reclamation, and termination are manual processes — which they are in most companies — it takes yet more IT resources during a time when IT is trying to focus more on strategic initiatives.

Difficult to reduce costs

Every year, IT tries to reduce costs, such as by talking to the supported lines of business to get a handle on what applications and how many licenses will be needed. For the most part, this is in the CIO's scope of responsibility.

However, with complexities like shadow IT and unplanned demand for technology, these savings seem to net out. Every budget cycle, the CIO is having the same conversation with the CFO: "We saved X dollars lowering network infrastructure costs and killing these applications, but on the other side the unexpected demand seemed to eat up our cost savings. So, bottom line, we didn't really see any savings at the end of the year."

Desired state #1: Visibility into cost triggers

Daniel Hartert shares his dream that every technology-related element that leads to a cost item in the overall IT budget creates a trigger that raises visibility. Who generated the cost? Who approved the cost? What are the ongoing operating costs for that element?

This would be a much better environment for the CIO to manage and would bring them closer to knowing precisely what is in their total technology portfolio, why it's needed, when it's predicted to fail, and what the cost of the element is over its lifecycle from planning to purchase to EOL.

Depleted Workforce Experiences

We talk a lot about how the CIO needs to deliver exceptional customer experiences, but the CIO also cares about delivering exceptional IT employee experiences. It's not an uncommon complaint from recent graduates who spent $100K to $200K to get a university IT degree only to find their IT job painfully boring because they spend their days closing out tickets, resetting user passwords, and updating Active Directory. Even though they get paid well, they'd rather be implementing new systems in the Performance and Transformation Zones that impact the business.

Job flight

Assigning boring or overly simple tasks is also a recipe for losing employees. From an emotional perspective, being bored at work is tiring, even if very little actual work was completed.[7] This is because employees that are unhappy spend considerable energy suppressing their emotions and can become both physically and emotionally exhausted. And beyond being bored, employees who can't get correct or timely information to properly support customers can also become extremely frustrated, especially when that means responding to disgruntled customers.

Unhappiness at work can come from other places as well, of course. The ability to keep employees productive by provisioning and maintaining the technologies they rely on also contributes to employee satisfaction. We all know the frustration we feel when the network is too slow or an application you need to meet a deadline is down.

Companies with employees experiencing low job satisfaction and engagement take the risk of those employees questioning why they work there. For many employees, they answer this question by moving on, especially in a hot job market when the next job is only a few applications away.

Losing an employee can have a drastic effect on team morale and lead to other dissatisfied employees asking themselves the same question. Before you know it, multiple employees have hit the offboarding process.

Refilling positions also adds cost to the CIO's budget, as it takes 24 days on average to fill a job, costing employers an average of $4,000 per hire[8] This doesn't include the cost required to train the new employees and payroll effectively lost as they come up to speed.

Tedious work is a recipe for errors

It's well understood that manual processes are more error-prone than automated ones. When you add repetitive and tedious to manual jobs, the chance for human errors just gets greater.

For instance, the exponential rise in the number and type of devices, users, and things connected to the network means networks are getting more complex and deployments more massive. Yet today, 95% of network changes are done manually, resulting in operational costs that are 2 to 3 times higher than the cost of the network. These manual processes also lead to configuration errors and inconsistencies in the network, for instance as disengaged employees make "fat finger" typing errors.

Desired state #2: KPI on workforce experiences

What the next CIO needs is a way to measure both customer and workforce experiences, for instance, as it relates to executing the processes in support of the business. Identifying areas of dissatisfaction may improve not only surface inefficiencies in the organization (like bored employees doing tedious tasks), but also enable a more proactive fix of issues that can help reduce employee turnover.

Expanded Audit and Compliance Risks

Components of risk management — which includes various security, compliance and audit frameworks, specifications, and processes — have one information element in common: they all rely on timely and accurate inventory and operational state data across an enterprise's technology portfolio. For example, the popular CIS 18 Critical Security Controls framework[9] specifies inventory and control of hardware and software assets, as well as configuration management, access control management, infrastructure management, and protection mechanisms. These same specifications appear in numerous other security and risk management frameworks.

Considering that, a foundational challenge of CIO, CISO and staff is having a lot of important user, hardware, application, network infrastructure and cloud infrastructure data scattered among a multitude of point management tools managed by different parts of the organization including the IT operations, security, compliance and audit teams.

Inefficient to satisfy audit requests

This data fragmentation makes it difficult for the IT organization to consolidate and normalize the data to provide to auditors proof the organization is successfully meeting internal, industry and regulated compliance obligations.

Talking to one CIO, he said it routinely took his team several days to respond to an audit request, effectively needing to take these four steps:

1. Access different systems to pull data
2. Upload this data into spreadsheets or some form of data lake
3. Analyze, edit, and normalize the data
4. Send the consolidated data to the auditors

In some cases, the IT team would create reports from the data; however, in most cases, they didn't complete a large analysis (often given a lack of time), leaving this up to the auditors to finish the work to fulfill the assessment. Regardless, it took a lot of time and energy from the IT teams to gather and normalize this data in support of an audit request. Unfortunately, the teams spent much of their time tactically collecting and preparing the data, versus adding more value to the organization by strategically analyzing it.

Compounding the challenge, any exception or anomaly in the data frequently created delays in the audit process, requiring the IT team to devote even more time diving back into those point tools to remediate any discovered issues. Of course, retrieving data through manual processes increased the risk of anomalies due to human error.

Further compounding the difficulty, this CIO's organization didn't have a standardized and repeatable process to respond to and satisfy audit inquiries. Consequently, it took even more time and effort to answer ad hoc auditor inquiries.

Again, the foundational challenge this CIO's team faced was distributed and fragmented data amongst numerous point management tools and systems. Combine this with manual to semi-automated processes that weren't highly standardized, and the result was that audits consumed a lot of time, resources, and cost from already stretched teams and budgets. And if an audit ended up being incomplete or inaccurate, the organization was at risk of fines and their reputation being diminished.

From talking to other CIOs, this situation is more the norm than an exception across the industry.

Difficult for CIOs to manage processes

Likewise, it's difficult for the CIO and senior management to know that the proper processes are in place and running correctly and efficiently. For instance, a CIO must often sign documents containing hundreds of pages describing the process landscape to satisfy one or more audit / compliance mandates. The CIO must trust the organization that what is described in the documents is in fact true.

Furthermore, the authors of these process documents typically come from the risk compliance department and are not the operating managers. This creates challenges between those who define the processes and those who must adhere to them (and be able to prove it).

Oftentimes, processes defined in audit and compliance readiness and supporting evidence documentation are not followed, either because people get too busy, the processes are incomplete, or there are too many changes and exceptions given that the real world is more complex than envisioned in the process document. These are the culprits behind many generated errors and omissions that delay audits and increase related expenditures (and fines).

In essence, it can be very difficult for the CIO to feel confident that the processes implemented by the IT organization are being performed adequately with auditable controls relevant to regulatory, industry and internal specifications. In addition, when the processes are manual, it's cumbersome to efficiently show proof that the IT organization is doing what it said it would do to meet those specifications.

Increased Security Risks

Every business in every industry is vulnerable to cyberattacks, as evidenced by the many news reports about companies being attacked, from healthcare providers to pipeline operators to meat packers. The outages caused by such attacks cost businesses more than time and money for investigation and recovery actions. Any potential data breach publicity can also damage company reputations that took years to build but can be tarnished in just a few hours.

According to a 2020 Cisco survey[10] of 2,800 CISOs and IT decision makers:

- 57% of respondents said the time needed to detect threats was a critical KPI for their business
- 81% reported that a challenge they faced was orchestration between security-point products
- 77% said they planned to automate more actions

The driving force here is the simple fact that data used by security operations today are mostly disjointed.

- The task of **preventing threats and investigating incidents is very complex** because companies utilize several IT management and security-point tools that share limited intelligence and context between them. This forces teams with the job of proactively remediating exposures and fighting cyberattacks to use multiple consoles and analyze varying data. While a security operator may have an IP or MAC address and other system level details, and perhaps network level details, it is common for the operators to have to further explore lifecycle as well as location, department, and owner details, which are often incorrect and outdated.

- The use of **point tools creates siloed visibility** with dispersed and fragmented data and other protections across a company's infrastructure and down to endpoints, manifesting as missed policy violations, anomalies and exposures, as well as longer dwell time and mean time to resolution following a successful breach.

- These factors make **security operations highly inefficient**, as teams are forced to manually coordinate multiple groups across operational domains and pull data from assorted sources to complete investigations and recovery actions. This leads to business disruption and employee burnout.

Regardless of whether the CISO reports directly to the CIO or is a peer, the CIO needs to help deliver these three things:

- **Faster detection**: Accelerate time to discover threats, anomalies and issues often leveraging security tools and frameworks.

- **Simple investigation**: Accelerate time to investigate using contextual information and other attack surface intelligence — for data, insight and more — in a single view.

- **Confident response**: Accelerate time to remediate with automated workflows where possible to reduce effort and strengthen security postures.

The simple fact is that the CIO, CISO and staff will struggle to deliver on the above points if they're running an IT organization that utilizes manual processes to assess scattered and fragmented data from siloed point products, no matter how strong those security and management tools are.

Indeed, processes that respond to security incidences is an area where workflows aren't automated by any stretch of the imagination, according to Randall Spratt, former CIO/CTO of McKesson Corporation. The only part that may be automated is perhaps isolating the incident by point security tools, but this is the smallest part of incident response, protecting against further harm. However, if there has been an incident, a lot of work is required and is mostly done through spreadsheets and policies.

Security frameworks and tools help, but you still need the data

Of course, there are many security frameworks that help companies know what they need to know to improve their security posture and reduce incident risk; however, the problem for the CIO, CISO and their staff is how to obtain this information and manage it on a day-to-day basis.

For example, many security teams rely on Security Information Event Management (SIEM) tooling and other data platforms to capture operational event logs from a variety of sources producing this log data. SIEM correlates this information to monitor for key behaviors, anomalies, and issues, and helps respond to issues and supports investigations. Unfortunately, these tools often do not incorporate the lifecycle management, ownership, location, and other context across the complete technology portfolio — crucial information for the security, audit, and compliance teams to optimize detection, investigation, and remediation efficacy.

Here again, this is very challenging using point tools, as the relevant data is scattered and fragmented.

Multi-cloud compliance and security example

For example, according to one survey, 90% of companies today run applications and workloads in multiple cloud environments.[11] This means the IT organization is likely responsible for numerous resources spinning up and down in various infrastructures, ranging from their own on-premises environments to public clouds like AWS and Microsoft Azure.

Another survey found that 61% of these companies use separate tools to manage their on-premises and cloud security postures, and 84% of organizations lack unified cloud visibility.[12] This is an indication that IT and security staff are using a variety of tools to identify unaccounted for, unmanaged, or at-risk resources in each of their cloud environments. Without centralized and actionable control data, organizations will continue to deal with inefficiencies to preempt exposures, delayed responses to issues, and inaccurate tracking of risk mitigation.

As a simple example, often different groups within a company spin up cloud resources for a variety of reasons such as in support of specific marketing campaigns, application development or Shadow IT efforts. When this work is completed, it is not guaranteed that these owners will take down these resources. Or if an employee leaves the company, there is no guarantee that that employee will transfer responsibility to another.

How does the CIO have informed certainty that all cloud resources the IT organization is responsible for, or are operating in connection to the business, are being properly managed and secured? If a resource is left running unattended and unused, not only does this represent wasted spend, but there could be vulnerability exposures and compliance violations that can lead to a data breach or at least an audit issue.

Too often, finding answers to these questions requires tedious, manual work. Typically, the IT organization lacks not only centralized visibility, but more so the process automation to proactively and efficiently highlight assets of concern to enable people to make informed decisions to reduce potential compliance and security risks.

So, let's say the CIO, CISO or auditor wants to know if there are any cloud resources that are unaccounted for or running outside of policy and therefore pose a potential security risk. And let's say this task is given to someone from the security team to complete.

Unfortunately, that person likely can't answer this question directly if using a point security management tool, because that person is operating with incomplete information. To start, that security tool might not keep track of the owner of a cloud resource, which may be stored in a cloud management tool. Let's say this person takes the extra manual step to inspect the cloud management tool for cloud resource owners and discovers that there are indeed owners listed for all cloud resources. Unfortunately, without parallel visibility into the HRM system, that person may not see that some of the listed owners of cloud resources are no longer employees at the company, making the resources unmanaged and potentially vulnerable to cyberattacks.

And when a company has hundreds to thousands of cloud resources running in different clouds, it can become an extremely intensive task to manually surf or attempt to manually integrate data from point tools in support of a simple use case. One example is making sure all cloud resources have owners that are active employees of the company and that these cloud resources are being actively managed. This is one reason why this is not often done consistently, and rarely efficiently. On the other hand, employees and contractors leave companies all the time, meaning that the risk of cloud resources not having owners and being found not appropriately managed and secure becomes a very real concern.

Desired state #3: Document processes with workflows defined in software

The next CIO needs processes not to be documented in multiple, massive, and irregularly updated PDF documents (or worse, not documented at all), but rather documented as digitally defined workflows that address security, compliance, and audit requisites. Workflows defined and executed in software can both be automated and logged to always prove they were completed correctly. In addition, workflows defined in software can better represent the work that needs to be and can be done versus text in a PDF that can deviate from reality. And since workflows defined in software are self-describing and run in production, they are more likely to be kept up to date.

Desired state #4: Leverage software-defined workflows to enable audit readiness

The next CIO's IT team could then utilize software-defined workflows to automate accessing data from the point tools, transferring this data into a centralized data warehouse, and analyzing and normalizing the data to be sent to auditors. Essentially, this would create a standardized, end-to-end ET process that utilized automation to quickly consolidate data for auditors. And by not requiring manual tasks, the data should be more accurate, reducing the risk of audit delays and violations.

Additionally, these software-defined workflows could include notifications triggered to reduce audit, compliance, and security risks. For example, how much time and risk would be reduced through an "unmanaged to managed cloud instance" workflow? In a multi-cloud environment, a notification could be triggered if a cloud instance has not been accessed for more than sixty days, or if there is no longer an active employee assigned to a resource as owner. In response to the notification, the CIO's team could then deactivate an at-risk, non-business critical or out-of-production cloud resource, notify the department associated with the resource, and wait for an employee or department to claim ownership to justify spinning the resource back up.

To be clear, these software-defined workflows would not be used to pass an audit. Rather, they would enable the IT organization to be audit-ready and facilitate compliance validation in a more automated and consistent way.

Desired state #5: Automated processes to gather and standardize data

The next CIO's IT team needs standardized and automated processes built on software-defined workflows to gather and analyze relevant data and monitor attributes within objects of that data to satisfy audit processes. To improve a company's security posture, an automated process could be deployed to integrate non-security point tool data with SIEM-captured data to provide greater context and operational oversight. This would help the CIO, CISO and their teams to more proactively identify and more efficiently respond to security threats and issues.

In addition, this more holistic view of previously fragmented data would provide more accurate, timely and complete data to support compliance and auditing tasks. Similarly, having access to this more holistic view would help the CIO feel more confident that the IT organization is meeting its compliance obligations.

Unproven Value Creation

Ultimately, the CIO and the IT organization must be seen as frontline value creators to the organization, versus unavoidable second-class cost centers. When IT rolls out a new application with the intention of creating value for the business, the CIO needs to know if value has in fact been created and by how much, not only to better lead the organization, but also to report the value IT delivered back to the rest of the business.

Delivering value through optimal process delivery timelines

Value can also be created by completing processes within a timeframe that optimizes the organization's operational efficiency. For instance, delivering an onboarding process to the business where every new hire has everything needed to be productive from the moment they start work helps to maximize value attained from this expensive headcount.

Ideally for each process, a CIO knows the precise cost and how long it takes to complete. This combination of timeframe and spend maximizes value creation to the business. Manual processes, of course, make doing this impossible.

Leveraging APIs to measure experience value creation

Measuring customer and employee experiences provides a place to measure value delivery by IT and makes a lot of sense because companies have increasingly more direct connections between their products and services and the end customer.

For instance, SaaS vendors routinely offer their APIs to report product adoption to themselves as well as to their customers, motivated by the desire to sell renewals. This data is a good starting point for IT. For instance, it can reveal if budget is being wasted on unused licenses, but it doesn't articulate the value the customer and employees receive from the use of these products.

The trend in the industry for IT service providers to provide Experience Level Agreements (XLAs) might provide some insights. Routinely, IT service providers delivered their services around Service Level Agreements (SLAs), which primarily focus on Quality of Service and measures metrics such as availability, capacity, and reliability. However, a shortcoming of these metrics is they don't measure the end-to-end experience of the users of the technology. For instance, was an application slow during a critical period? Was the network slow because of misconfiguration of the SD-WAN?

XLAs put added focus on the Quality of Experience by measuring how satisfied the user is with the IT service or the process of obtaining the service, ensuring that all service interactions and touchpoints are considered when defining whether the service meets the agreed performance level. Although XLAs normally focus on customer experience, for our purpose we expand this definition to include employee experiences as well.

Typically, user feedback is the primary measure used in XLAs. This is why most digital service providers will ask for a 1 to 5-star rating at the end of each request or issue as a way to measure customer satisfaction. Another common element used to measure XLAs is the Net Promoter Score, which measures how likely the customer would recommend the service to another. And another metric, the Customer Effort Score, measures the complexity of a process the customer underwent in the provisioning of a service.

All these approaches are helpful, but at the end of the day they tend to be subjective. In addition, feedback from users can be difficult to collect.

A more sophisticated approach would be to track the application's API calls, understanding what each user is doing on the application. This approach might hold more promise to surface insights given modern applications are increasingly built from a collection of web services calls. The CIO would then be in a much better position to cut programs that aren't delivering enough value for the budget spent to keep them going.

The bad news is that this creates an enormous amount of data, impossible to sort through manually. The good news is that the industry has learned how to extract value from enormous amounts of data through AI/ML. The challenge with AI/ML has always been access to the data to train the AI/ML engines.

So why not figure out how to leverage the APIs from an organization's technology portfolio to give quantifiable measurements to measure experience levels for all the services delivered by IT to the business?

Desired state #6: AI/ML informing KPIs that measure actual IT value creation

Where are the KPIs that measure ET process performance to value created and delivered that improved business outcomes? The next CIO needs to decide what KPIs are needed to measure value creation for the various ET processes, and then leverage AI/ML to inform these KPIs.

Furthermore, the next CIO would have one dashboard to view the updated KPIs to always have a pulse on how much value the IT organization is providing back to the business, both in terms of customer and employee experiences mapped to defined business outcome requirements.

Finally, with smarter processes comes the opportunity for proactive versus reactive responses. Wouldn't it be better to be proactive to improve an employee's experiences and satisfaction versus reactive, perhaps after the employee has already checked out and accepted a new position with a competitive firm?

If the next CIO could solve many of these problems, how would that impact available budget to fund Transformation and Performance Zone initiatives? Let's take a look.

Chapter Summary

Key takeaways
Running an IT organization with immature and manual processes — in combination with having data about the enterprise technology

scattered and fragmented across a multitude of siloed tools — creates several challenges for the CIO.

- Leads to overinvesting in IT infrastructure, unoptimized endpoint lifecycle management, and challenges tracking technology inventory and spend resulting from Shadow IT. This makes it difficult for the CIO to reduce costs because successes in cost reduction seem to be frequently nullified by unexpected demand for more technology spend.
- Leads to depleted workforce experiences, as manual processes can be tedious and boring, leading to increased human errors and job flight.
- Expands audit and compliance risks, as manual and poorly defined processes relying on scattered and fragmented data make it highly inefficient for IT teams to respond to audit requests. This also makes it harder for the CIO to ensure that processes are being followed correctly.
- Expands security risks, as a lack of automation and a holistic view of the data relevant to prevent, detect, respond, and remediate security threats makes it difficult to strengthen security postures and respond to cyberattacks quickly.
- Makes it difficult for the CIO to articulate IT's value creation back to the company, reinforcing the misperception that IT is only a cost center.

Key questions
- Do you operate immature, manual processes?
    - Are you overinvesting in IT infrastructure?
    - Are you struggling to manage your technology lifecycle buys?
    - Are you having to respond to expensive fire drill demands for technology purchases?

- o Is Shadow IT making it difficult to keep an accurate technology inventory and budget?
- o Do you lack visibility into cost triggers?
- Are you challenged to deliver exceptional employee experiences?
  - o Do you struggle with employee dissatisfaction and job flight?
  - o Do any of your employees tend to an abundance of tedious manual tasks
  - o Do you have issues with human error impacting your ET processes?
  - o Do you lack KPIs that measure your employee experiences?
- Do you have audit and compliance risks?
  - o Is it difficult for auditors to perform audits?
  - o Is it difficult for CIOs to govern processes?
  - o Are your workflows defined in software so they can be more easily described and automated?
- Do you have security risks compounded by immature ET processes?
  - o Do you have disjointed security operations?
  - o Do you lack holistic visibility into your security posture?
- Can you prove your value creation to the company?
  - o Are you not delivering value through optimal process delivery timelines?
  - o Are you not leveraging APIs to measure experience value creation?
  - o Are you not leveraging AI/ML to drive KPIs that measure actual IT value creation?
- How does IT articulate its value creation back to the organization?

Chapter 4: Where's the Money?

What if ET processes could be better defined, automated and optimized? How much potential budget could IT free up? To understand this, the CIO needs a strong partnership with finance.

On one hand, finance could help the CIO's office categorize IT budget spend by Zones to Win. In parallel, the CIO would do well to present a plan to reduce Productivity Zone budget spend to fund transformation initiatives, without the need to increase IT's overall budget.

Let's do a little back of the envelope calculation to see why (borrowing heavily from Toby Redshaw's experienced insights).

Finding budget

Figure 4-1: Mythical status quo IT budget

Let's say you're the CIO of a major enterprise with a one-billion-dollar IT budget.

Let's also say you need to find budget to fund some disruptive new what about initiatives like digital transformation.

Looking at Figure 4-1, let's say that half of your budget – $500M – is spent keeping the lights on with Productivity Zone projects. Now, many CIOs of large enterprises might look at that 50% assumption and laugh, claiming that the number is more likely 70–80% of their budget. In Edward Wustenhoff's (current Vice President Infrastructure and Platform Operations at Juniper Networks) experience, this number is more like 90%. If these numbers are true, this only reinforces my point.

Let's also assume that 40% of your budget is spent on initiatives that impact the business and live in the Performance Zone. (This might be generous, as many IT organizations likely only spend 20–30% of their budget on business impact initiatives).

Finally, we'll assume that a full 10% of IT budget is already being spent on Transformation Zone projects like migrating to the cloud or implementing an IOT strategy.

Take Productivity Zone budget down

Over a three-year period, let's say the IT budget reduces by 10% down to $900 million due to natural savings and the IT organization implements some level of continuous improvement.

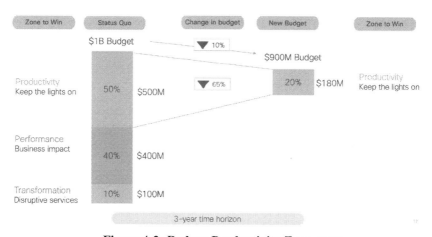

Figure 4-2: Reduce Productivity Zone costs

Given this backdrop, what would happen if over this three-year horizon, the CIO could keep the lights on using only 20% of the overall IT budget, as shown in Figure 4-2? This would mean taking the $500M spent funding Productivity Zone work down to $180M for a 65% spend reduction.

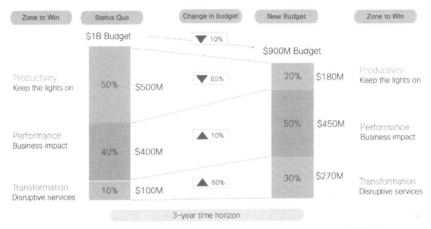

Figure 4-3: Increasing Transformation budget by $170M

Take Transformation and Performance Zone budgets up

Out of an overall $900 million IT budget, this savings would free up $720M to fund Performance and Transformation Zone initiatives (as shown in Figure 4-3) — an increase of $270M over the status quo spend of $400M for Performance Zone work plus $100M for Transformation Zone work.

Further, if the CIO increases Performance Zone budget by roughly 10% to $450M, this will provide a huge $170M budget increase to fund Transformation Zone initiatives, like digital transformation.

Gain further value improving operating efficiencies

But wait, there's more! Let's talk about improved operational efficiencies.

Being able to reduce budget to cover Productivity Zone work by this much this would likely be the result of running a smarter organization. This cleaned-up environment would likely be at least 20% more effective per unit dollar spent because there's less ambiguity and the environment is instrumented to enable higher levels of efficiency.

So, now that $450M budget allocated for Performance Zone work is probably bringing 1.2 times $450M (or $540M) worth of value to impact the business. This is more like a 35% increase from the original $400M budget. Likewise, the $180M spent on transformation is likely bringing over $210M worth of budget value to the organization.

Agile and fast beats big and slow

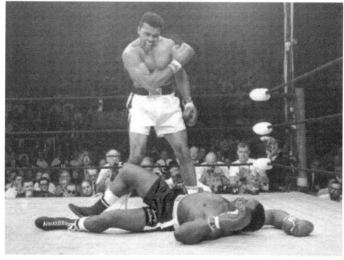

Neil Leifer for SI

We all know that agile and fast beats big and slow. Mohammed Ali standing over Sonny Liston less than 2 minutes into the first round of their second heavyweight championship fight paints the picture. Ali, who was known for his fast but light punches, made the hard-hitting heavyweight champion Liston look, well...awkward.

In fact, if you're agile this should be marketed as a competitive advantage and positioned as part of a company's offering. And the level of agility is much higher when it's built into the processes that ultimately deliver the products and services the company delivers, both to the external market and internally to help run the business more optimally.

So, what's stopping an incumbent IT organization — no doubt filled with a staff of hard-hitting champions — from increasing their agility and improving operating efficiencies to the order of reducing Productivity Zone work by 65%?

Perhaps it has something to do with how well IT is being managed.

Chapter Summary

Key takeaways
- If the CIO can significantly reduce costs in Productivity Zone work by making enterprise technology processes more efficient, this could free up budget to fund Transformation Zone initiatives, as well as increase funding for business-impacting Performance Zone projects.

Key questions
- At a high level, what percentage of IT budget (and how much money is that?) is spent funding Productivity, Performance, Incubator and Transformation Zone projects?

- Where else is money spent on processes used in the Productivity Zone?
- For each process used in the Productivity Zone, what inefficiencies from Chapter 3 are leading to overspending? What is the calculation of: If this problem were reduced by X%, then we could save $Y?

Chapter 5: IT is Not Managed Properly

According to Toby Redshaw — CEO of Verus Advisory and former global Top 50 CIO (InformationWeek) and Top 25 CTO (Infoworld) — neither a tech-savvy nor a business-intensive approach to IT will be sufficient as we enter the 4th Industrial Revolution and face the concurrent waves of change. A new model that integrates business, technology and an architecture for agility/cycle time is the correct competitive framework going forward.

Redshaw elaborates that today's IT organization is like a Russian Matryoshka doll. Given that digitization is a fundamental driver for the 4th Industrial Resolution, the outside of the doll represents the immense importance of the CIO and IT organization. However, when you look inside, you realize IT is not managed properly.

Of course, saying IT is "not managed properly" is a big statement to make, especially in a book targeting the very people that manage IT organizations.

Follow along with Redshaw's work to hear him back up this statement in his upcoming book (https://www.verusedge.com/the-book). On one hand, Redshaw will help to demystify the future of tech. On the other, he will provide pragmatic advice on how to manage through the upcoming change impacting any technology-reliant company — which is essentially every company.

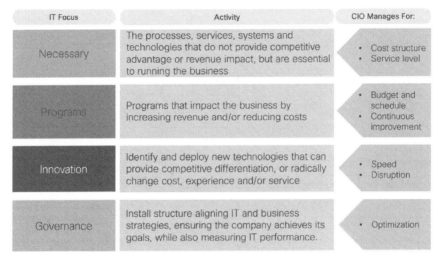

IT Focus	Activity	CIO Manages For:
Necessary	The processes, services, systems and technologies that do not provide competitive advantage or revenue impact, but are essential to running the business	• Cost structure • Service level
Programs	Programs that impact the business by increasing revenue and/or reducing costs	• Budget and schedule • Continuous improvement
Innovation	Identify and deploy new technologies that can provide competitive differentiation, or radically change cost, experience and/or service	• Speed • Disruption
Governance	Install structure aligning IT and business strategies, ensuring the company achieves its goals, while also measuring IT performance.	• Optimization

Figure 5-1: What the CIO manages for

In the meantime, here are a few points drawn from Redshaw's extensive insights learned over several years as a CIO and CTO. These insights are organized into four IT focus areas: Necessary, Programs, Innovation and Governance (shown in Figure 5-1).

Manage the Necessary

One thing the CIO worries about is the necessary processes, services, systems, and technologies that do not have direct impact into revenue or competition. These fall squarely into the Productivity Zone.

They include services shared by the business, such as in the areas of finance, accounting, security, email, order processing and technical support. These things only have an impact on the business when they fail. No one cares about the General Ledger competitively, but a failure during a quarter end is a catastrophe and can even impact stock price when the market postulates why reporting was late.

Manage for cost structure and service level

In this category, the CIO needs to manage for both cost structure and service level — a difficult balancing act for sure.

For example, to help reduce costs, this has become an area rife with outsourcing to managed service providers where the CIO can routinely publish new RFPs to solicit bids that squeeze out as much cost as possible. That said, this can also be a ticking time bomb if the service level component suffers: The email service IT delivers to the business may not provide market differentiation, but it can sure cripple the organization if it stops working.

On that note, managing for cost versus cost structure are different. As Redshaw muses: "Do you want the factory where they've cut costs by 30%, or do you want the factory where you get 30% more unit output for the same cost?"

You want the second one, of course, because you don't know what they did to cut costs by 30%. Did they remove safety measures? Are they now using less expensive, but much less reliable equipment, increasing the risk of availability issues going forward?

Managing for cost structure is more complicated than just managing for cost, as it puts more emphasis on process and quality of the output whether it be in delivering widgets or an IT service.

When the CIO manages for service level, he or she is responsible for delivering on three value propositions:[13]
1. Improved efficiencies ("doing things right")
2. Improved effectiveness ("doing the right things")
3. Regulatory compliance

Unfortunately, many CIOs take their eye off leading in these areas. It's one thing to outsource the payroll system for less cost, but it's very problematic to the organization if the payroll system is slow, down, or never quite works right.

An infamous example of this is when the health department of Queensland hired IBM to develop an application for administering payroll. IBM estimated the cost to be roughly $6M. Soon after placing this estimate, IBM realized that the job was going to be much more work than they had thought. In the end, the project came in at $1.2B and never worked properly.

There was not a lot of cost savings or optimal service delivered on that initiative.

Current CIO and Senior Vice President of Global IT at Novo Nordisk, Anders Romare, shares a somewhat related example. Before the pandemic, his IT organization needed to procure laptops for employees. His team was able to reduce costs significantly by entering into an agreement with a particular low-cost vendor. However, after the pandemic hit and employees were suddenly working from home using their laptops the entire day, the laptops began to overheat and have battery life issues. So, suddenly a decision that was more optimized for cost ended up being suboptimal in terms of service level delivered to employees.

That said, managing foundationally for service level requires access to data. For instance, improving efficiency and effectiveness requires knowing, at a minimum, the benchmark of where things were before and where things are now because of our efforts. In terms of meeting compliance obligations, the CIO needs access to the data that confirms processes are being followed properly (or not).

Manage Programs

The second category of things a CIO worries about includes programs that do impact the business, for instance by helping to drive revenue or reduce costs. These programs typically live in the Performance Zone, such as upgrading existing retail point-of-sale devices or modernizing customer-facing applications to cloud-native architectures to deliver better user experiences.

Manage for budget and schedule

When running these programs, the CIO typically manages for budget and schedule, monitoring if the work is progressing according to plan while minimizing spend. If a program is to modernize point-of-sale devices in the retail stores, it's ultimately the CIO who is responsible for the IT organization meeting its delivery and quality commitments to the business while keeping within budget. Given these programs typically have an assigned project manager, the CIO should be kept well informed on program status.

Manage for continuous improvement

However, one area often missed by the CIO is continuous improvement of those programs. It's rare, for instance, for anyone within the IT organization to review a program status twelve months after implementation to validate if it is meeting the objectives set forth when gathering approvals to do the work.

One reason for this is the difficulty of accessing the appropriate data. The CIO has no application (currently) where a click of a button provides KPIs for implemented projects, including comparing actual numbers to the ones used to justify doing the project in the first place. Rather, to get these numbers, the CIO would likely need to reassign resources to manually try to figure out status and ways the implemented projects could be improved. Never going to happen. And it doesn't.

Manage Innovation

Third we come to innovation, which includes initiatives that sit in the Innovation Zone. With innovations, a CIO is tested on how well they can see over the horizon to what technology may eventually impact the business. These technologies generally fall into one of four buckets.

- **Irrelevant**: The technology was looked at, but it has nothing to do with the CIO's company's business. For instance, a new 3D-printing technology could be very cool technically, but not be impactful to the business.
- **Park**: The technology might be interesting. We'll keep an eye on it and peek at it again every six months or so to decide whether it should be moved to one of the other categories: irrelevant, sandbox or disruptive.
- **Sandbox**: The technology could provide transformative advantages, which could put our business at a disadvantage if our competitors are adopting it. The CIO puts these technologies into a sandbox and has a team play with it to confirm if it's truly transformative or not.
- **Disruptive**: This decision is easy, as it's clear to everyone in the room that this technology is transformative; we need to adopt this technology now.

Managing for speed

As such, with innovations, the CIO needs to manage for speed. How quickly can the potentially disruptive technology be vetted and then brought into the Transformation Zone and ultimately put into production to improve the business outcomes?

One could argue that this is one of the CIO's most enjoyable jobs. What techie at heart doesn't love checking out new technology? It's sexy, cool, fun. It's where you can dream.

But the pace of today's business can slap you back into reality like being dropped into a Formula One car in the middle of a race: hit the accelerator or perish into last place.

Put another way, once a disruptive technology is discovered and vetted, the CIO has the enormous task of managing the process of driving enterprise-wide change, so the innovation is adopted by the organization.

Managing for disruption

As the CIO, your Formula One car is your team's ability to steer through disruptive forces at every turn and straight away. The thing is, before digital transformation was on the CIO's agenda, the CIO was already navigating a field full of other what about disruptors.

For instance, migrating to the cloud is on every CIOs to do list. Which applications do we *lift-and-shift* (rehost) to the cloud and which do we refactor into a cloud-native architecture? Which do we run in a hybrid cloud architecture or keep on-premises due to security or compliance reasons?

The pandemic introduced practically every organization to the need to support hybrid work for its employees. How do we secure applications and data no matter where employees work, while providing the application user experience they need to be productive?

With the advent of cloud computing, many CIOs are running programs to consolidate their data centers. But now the CIO must also deal with effectively securing thousands of data centers, as organizations have proprietary data in SaaS applications hosted on infrastructure that the CIO does not control, while each remote employee in a way acts as its own data center with in-home routers, VPNs, compute and storage.

IOT technologies extend the landscape the CIO manages because data is streaming in from thousands of endpoint sensors. How does the business gain insights from all this data?

The point is, before digital transformation came along, the CIO was already managing several disruptive services that held the promise of transforming the business and were crucial to the business remaining competitive in the market. Not to mention, the CIO knows the organization's competitors are in their own race cars also hitting the accelerator trying to create value from the same disruptive technologies.

Okay, so the race is on. But why do we have CIOs steering and managing with one hand tied behind their backs? For instance, where's the data to help the CIO and staff manage all the processes required to deliver services back to the business?

Manage Governance

On that note, the fourth category of focus is governance. Key to governance is giving the CIO and staff access to timely data, ideally turned into insights, to help optimally manage the other three categories.

Manage for optimization

For service levels delivered back to the organization, what are the KPIs and notifications to ensure that the performance and availability delivered to the business are aligned with SLA commitments?

How fast are new, potentially disruptive technologies being evaluated to ensure R&D isn't spending too much time playing around with technologies they've become enamored with but have no path to adding value to the business?

When new, innovative technologies are deployed, what is the governance for getting rid of the old stuff? Redshaw mentions that he asks that simple question at every company he's ever visited and, except for three, none had an explicit process for removing replaced technology. Consequently, you get applications and infrastructure that are past end of life still taking up power and data center space.

One CIO said that during a data center transformation project, moving from running their own data center to a co-location provider, they discovered that 24 percent of the running hosts had already been retired and were no longer in use. One example culprit to this was a line of business no longer running an application, but there being no process in place to tell the IT organization this small but important fact. As a result, the IT organization ends up still running the application, paying for infrastructure (and the power and cooling for that infrastructure) and backups.

To help with governance, in addition to having access to data, it's important the CIO have internal partners that can help. Finance is an ideal candidate, for instance, helping to manage cost, as Finance can put in spending controls, so technologies aren't purchased that don't meet the company's policies.

At one company, Redshaw asked the CIO how many instances of Salesforce they ran? The answer was none: they were an SAP shop. Then Redshaw went to procurement and asked the same question. The answer was they had 112 contracts for Salesforce in place. Each contract snuck in because it would add value to the group making the request, but the governance and spending controls weren't in place to at least flag that the request wasn't on the targeted architecture.

This begs an important side note. Frequently, a siloed group within the line of business or IT organization are doing things — like asking for a Salesforce contract — to perform their jobs as well as possible. They frequently don't have the optics to see the broader organizational picture to understand how their request might negatively impact the overall business. This is where it's crucial for the CIO and staff to have the processes in place, coupled with aligned metrics and notifications, to optimize workflows important to running the business.

Again, with all the technology we have available to us, why is the CIO still steering and managing the IT organization with one hand tied behind their back, not given access to the data-delivered insights needed to govern well?

The Role of the CIO

CIOs are extremely smart, knowledgeable, and capable people. There are many great reasons why the CIOs rose to the C-suite on the IT side of the organization.

However, CIOs are faced with an enormously challenging job, which I suspect most people in the rest of the organization don't fully appreciate. Unless you've had your hand in making technology work to solve real problems, it's hard to appreciate and value how tough this job is. As Anders Romare shares, "The CIO's job is especially challenging in that they must manage processes that span the entire value chain, while taking data input from a wide variety of technology sources. Add to this the fact that the CIO works with several stakeholders across the business."

Then to be tasked to do this on a continuous basis with customers willing to trash your brand if page load times aren't fast enough, along with threat actors just chomping at the bit to benefit from stealing your organization's data.

Then layer on other disruptive technologies that need to be integrated into the business, so everyone still has a job the next fiscal year and…phew, it's tough. It's complex.

According to former Safeway CIO David Ching, the role of the CIO is four-fold:
- **Operate**: This is the operational side of the house, with an IT focus on Productivity Zone, Necessary work and Performance Zone Programs, frequently governed by metrics and measurements in areas such as response times and meantime to repair.

- **Innovate**: This is where the CIO's value creation is mostly appreciated, bringing new technologies into the Incubation Zone and transitioning them to the Transformation Zone with a focus on improving the business status in the market.
- **Lead**: The third piece is on the leadership side where the CIO plays the role of executive officer, looking around the corners always with a focus on uplifting the IT organization, for instance driving strategy discussions.
- **Secure**: Some organizations have the CISO reporting directly to the CIO or as a peer. Regardless, the security technologies deployed by the business are managed by the IT organization, so one could include managing security as an additional role of the CIO.

Mr. Ching's perspective is that, especially in large enterprises, CIOs pretty much delegate the operations role to people under IT management that run the IT workloads more like a production shop. As such, CIOs may tend to give a slightly higher priority to their roles around innovating and leading.

So perhaps the role of operations isn't as sexy as innovating and leading, but at the end of the day, even the best in innovation and leadership can be stunted by suboptimal operations.

How can operations be as cool as innovating and leading? How do we enable the CIO to navigate and manage the race car with both hands firmly on the steering wheel, winning the race gaining value from disruptive technologies while optimally keeping the lights on?

Maybe we can learn a thing or two from modern factories.

Would You Run a Factory Like This?

Let's say you're the VP in charge of a factory that produces widgets. You're in a meeting with your company's CEO and CFO, who ask you these four fundamental questions:

- Can you show me the process and accompanied workflow deployed to build a widget?
- How long does it take a build a widget, including the standard deviation measured in seconds?
- What inventory of assets do you have on the floor that are used to build a widget?
- How much does it cost to build a widget?

What if your answers went something like this?

- I have no idea what process we use to build widgets, as we haven't written it down anywhere but I'm sure it's in the heads of some of our best employees.
- I have no idea how long it takes to manufacture a widget, but I suspect the standard deviation is more likely measured in days, not seconds.
- I don't have a list of the assets on the factory floor used to build widgets, but I could ask one of my employees to go figure it out for us. Give me a few days to answer that question, although I would need to pull that employee from his normal job.
- I have no idea what it costs to build a widget. Does finance know?

Then what if your CFO responded (in utter shock), "But we give you a billion-dollar budget to run the factory!"

And you reply, "Well, I can tell you widgets come out of the other end of our manufacturing line, but I really have no idea how it happens inside the factory. It's a bit of a black box to be honest."

At the end of a conversation like this you would no doubt be let go and put into an offboarding process (a process which may or may not be well-described, much less automated).

It doesn't take two years of business school to know that for a factory to run efficiently, all inputs, outputs and processes must be both well-defined and continuously measured to ensure quality, meet schedule commitments, and to find ways to improve processes. Manufacturing has processes in place to lower defects, improve quality, and reduce costs. Again, you don't have to be the smartest person in Business 101 to realize that lower cost and higher quality generally wins.

Leveraging a higher-level application

As the VP running a modern factory, you and your team would work closely with finance, likely leveraging an ERP application to help manage day-to-day business processes such as those found in accounting, procurement, project management, risk management and compliance, and supply chain operations.

Your ERP application would integrate point tools while automating and simplifying well-defined business processes, such as:
- *Market-to-cash* integrating business planning, the field sales plan, in-store execution, and integrated order management
- *Forecast-to-deploy* integrating demand forecasting, weekly demand and supply management, production, warehouse management and fleet maintenance
- *Procure-to-pay* integrating capital budget, vendor management, procurement management and accounts payable

You would have a top-down, holistic view of how long it takes and the cost to build each widget down to a standard deviation measured in seconds and pennies. You could click in to get details if needed through the ERP application. You and your team would know precisely what technology you have deployed to build widgets. You'd also be sent alerts when anything strays from the well-defined process base lines.

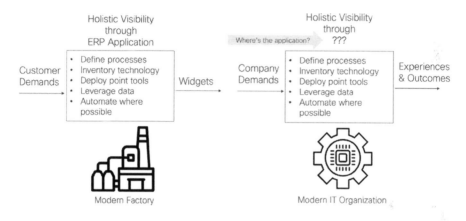

Figure 5-2: Parallels with operating a modern factory

As shown in Figure 5-2, there are many parallels between modern factory operation and what IT does. Whereas a modern factory utilizes customer demand to drive the manufacturing of widgets, IT takes in demands from the company to ultimately deliver experiences and business outcomes. Both organizations utilize defined processes and technology to get work done. Both utilize point tools and leverage data to make smarter decisions and automate whenever possible.

However, a factory leverages an application like ERP to provide high-level visibility and notifications. The CIO has no such application.

Instead, most enterprise IT teams today utilize a broad range of point technology management tools. To get an idea of what's going on across the entire technology portfolio, IT frequently needs to ping-pong between different tools and consoles. It goes without saying that this is not a very efficient way to go, especially as technology portfolios continue to grow in both size and complexity.

CIOs need their own application: ETM

Historically, an explosion in complexity has given rise to new categories of software applications that help simplify processes and drive efficiencies.

- ERP provides *procure-to-pay*, enabling organizations to manage day-to-day business activities more efficiently.
- CRM provides *lead-to-cash*, enabling sales organizations to manage company relationships and interactions more efficiently, leveraging large amounts of customer data.
- HCM provides *hire-to-fire*, enabling HR to manage traditional administrative functions more efficiently.

So, where's the CIO's application?

Sales has CRM
Lead-to-cash

Finance has ERP
Procure-to-pay

HR has HRM
Hire-to-fire

CIO needs ETM
Plan-to-EOL

Figure 5-3: CIOs need their own application: ETM

Enter the need for an Enterprise Technology Management (ETM) application. A higher-level application for the CIO working across the IT organization and the rest of the business.

An ETM application would connect with IT management point tools to exchange data, while automating and simplifying well-defined ET processes.

Demand forecasting example

Take the important process of demand forecasting, which includes budgeting replacements, refreshes and upgrades of technology already purchased by the company. Accurate forecasting around existing technology requires not only a precise inventory of what has already been purchased, but also data on how often the technology breaks down and needs replacing.

For example, a major retailer with hundreds of brick-and-mortar stores uses an ITSM application for its asset management tool. This retailer deploys thousands of POS devices in its stores. The retailer was struggling to accurately forecast when replacements would be needed, partially because store associates would often just set broken devices aside and keep on working, leaving the store with one less lane available for checkout.

The data that a POS device was broken was somewhere out there, but this data wasn't automatically getting back to the CIO's office to automatically trigger shipping a replacement. Other data was also not being leveraged, such as mean time to failure — data that could be critical to a seamless demand forecasting process that could order POS devices in anticipation of down devices in the field. This data would also be helpful when it came time to negotiate a new contract with the POS vendor.

The issue was that this retailer's point tools weren't integrated to automate data coming from broken devices in the field informing the demand forecasting and procurement processes. Instead, relatively expensive and slow manual labor was required to input the data into these processes. Meanwhile, longer replacement times meant increased risk of customer dissatisfaction due to slower checkouts (given fewer available checkout lanes).

Since the retailer's tools had open APIs, there was an option for the IT organization to fund a project to stitch a solution together. But that would require additional work on the already heavily burdened, lean IT team, taking them away from Performance and Transformation Zone projects. In addition, writing custom code can lead to more headaches and costs down the road as the code needs to be supported and updated as the point tools update with limited warning.

It's all very complex: Just trying to keep track of all the technology used in a modern enterprise — such as endpoints, networking, infrastructure, applications, and accessories — is complex enough. Then add the management processes that touch all this technology — from purchasing, deploying, and monitoring to securing and maintaining — and the complexity only grows.

Like ERP, an ETM application would provide a single interface to collect and organize data from the point products that hold the data about the various technologies to provide high-level visibility into the processes that touch the company's entire technology portfolio.

Let's take a deeper look at what an ETM application might look like.

Chapter Summary

Key takeaways

CIOs and IT have 4 focus areas and manage for six things:
1. Essentials Focus
    1. Manage for cost structure
    2. Manage for service level
2. Programs Focus
    3. Manage for budget and schedule
    4. Manage for continuous improvement
3. Innovation Focus
    5. Manage for speed
    6. Manage for disruption
4. Disruption Focus
    6. Manage for optimization

- We can learn from how a modern factory is run to inform how a CIO should run an IT organization.
- As the industry has applications in several verticals that provide top-down, holistic visibility (e.g., ERP, CRM and HRM), the CIO could benefit from having a similar application that provides the CIO with visibility into all the ET processes that touch technology. I call this application Enterprise Technology Management or ETM for short.

Key questions
- How would you rate yourself in the six CIO management areas from 1 to 10 (1 = poor, 10 = excellent)?
- What immature enterprise technology processes (if any) are making it more difficult for you to better manage in these six areas?

- If these enterprise technology processes were automated and optimized, how much better do you think you could rate your performance in these six areas?

Chapter 6: Introducing the ETM Application

If there was an application built to provide the CIO and his or her team with higher-level visibility and control into ET Management Processes that touch the entire enterprise technology portfolio, what would it look like?

Without trying to provide a formal product requirements document, let's take an architectural view of the potential ETM application.

ETM Application Architecture

Figure 6-1: ETM architecture foundation

Before we lay out the ETM architecture, let's start with the existing enterprise technology landscape, as seen in Figure 6-1. This landscape encompasses the company's enterprise technology portfolio, including endpoints, networking, infrastructure, applications and accessories.

Moving up the stack, the management of this technology is provided by the ETM application sitting on top of and interfacing with existing technology management tools.

It's imperative that an ETM application not require a "rip and replace" approach regarding a company's existing technology management point tools. Rather, it's best to integrate with whatever a company might already have in place. This would also provide flexibility for operators who can use, replace, or introduce new tools as needed.

To the right of the ETM application is the ET Process Maturity Framework, which provides a way to determine where each ET process stands on the maturity journey to becoming more efficient with more secure data. We'll discuss this in more detail in Chapter 7.

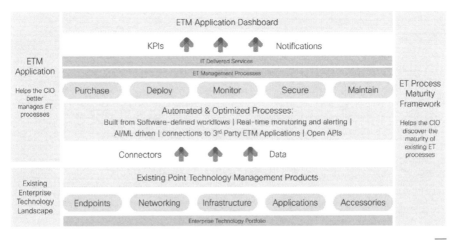

Figure 6-2: Full ETM application architecture

With that background, let's explore the ETM architecture — as seen in Figure 6-2 — in a bit more detail.

Integration and connectors

A paper[14] published by MuleSoft in collaboration with Deloitte Digital found that in 2020 large businesses had, on average, well over 900 applications of which only 28% were integrated. Further, the paper reported that 88% of respondents said, "Integration challenges continue to slow digital transformation initiatives." This was followed by risk management and compliance as the other top challenges to digital transformation. The paper further found that organizations are spending an average of $3.6M on custom integration labor, highlighting a significant drain on time and budgets.

An ETM application will need to help address this integration challenge by providing connectors between the siloed products and the ETM application itself. After all, if you think about it, a major motivation for integrating siloed IT management tools is to inform the automation of higher order processes. For instance, once one piece of technology completes a task, the process may require another piece of technology to start a subsequent task.

These ETM application connectors would effectively provide integrations with the existing landscape of technology tools already in use by the organization. These connectors would likely be in categories such as device management, identity management, and service desk management, for example:

- Assets such as Amazon EC2, Cisco Meraki and Tanium to assist with, for example, device and endpoint management

- Applications such as Webex, SmartSheet, Workday and Marketo to ensure, for example, that employee and data access privileges are being properly managed, and paid licenses are being optimally used
- Single sign on (SSO) such as Azure Active Directory, Google Workspace and OKTA to manage sign-on privileges
- Service desk such as ServiceNow, Zendesk and Salesforce to send notifications to the ETM dashboard, for instance, if tickets are not being managed properly

The ETM application provider would be responsible for providing these connectors, as they would be required for an ETM application deployment.

Data

Ultimately, these connectors provide a mechanism for data to be exchanged between the ETM application and the point products. There are several types of data, which provides a glimpse into how rich the potential data set can be.

- State data provides information about the state of the technology. Is it working or not? What version of firmware is it running? Where is the physical location of the technology?
- Access data includes access controls, which establish who has access to the technology and at what levels.
- Use data provides information about how the technology is being used, such as which parts of an application are being utilized the most.
- Nomic data[15] describes the external environment of the technology and may be more relevant to industry specific IOT devices that have sensors to collect external data like temperature, air quality or water flow.

The connectors would access this data through the point tools' open APIs.

Automated and optimized processes

The ETM application would then enable the creation of automated and optimized processes operating at the highest level of process maturity. These processes would consume and communicate data to the point products through the connectors, and have the following attributes:

- **Software-defined workflows**, likely through a low- or no-code drag-and-drop interface. This way not only can workflows be created by any part of the business, but they can be designed to comply with any compliance regulations and provide the data necessary to pass audits.
- **Real-time monitoring and alerting** to identify, visualize, assess, and trend condition changes and to trigger events, significantly reducing the time spent troubleshooting workflows, supporting auto-remediation workflows, and making it easier to optimize existing processes.
- **AI/ML driven to provide intelligent automation**, enabling proactive alerting of possible issues, as well as alert filtering to minimize false alerts and prioritize responses based on the importance to the organization, ultimately improving SLAs and XLAs.
- **Connectivity** to other customer ETM applications to take process automation and optimization to the next level, enabling collaboration with other customer ETM applications, sharing data and best practices.

- **Open APIs** into the ETM application database provides a way for third-party tools to directly access the "single source of truth" data managed by the ETM application.

These ETM application attributes would help the CIO and IT organization more efficiently automate and optimize the ET management processes, to ultimately deliver more value back to the business.

Regarding enabling connections to third-party ETM applications, Justin Mennen, CIO of Rite Aid, mentions: "Many IT organizations outsource some of their operational workloads to third-party service providers. Automated notifications regarding status of these managed processes would provide the CIO with broader visibility into the entire IT landscape."

Edward Wustenhoff points out a deeper challenge with outsourced service providers: "If an outsourced service provider brings in their external processes, tools and technologies into our environment, it can create the risk of confidential data being leaked. For instance, this might happen if they run their own agents on our servers that report back to their management platform." Wustenhoff wants to control the data and metrics, saying to "let the service provider define what metrics they need reported back to them to be successful to manage our technology. We control the data, and we control the metrics. And when we don't like or don't want to use the service provider anymore, we just turn it off."

Alain Brouhard, former CIO of Coca-Cola, brainstormed how connected ETM applications might also help with product quality, as many companies get their inputs from the same vendors. So even though Coca-Cola and Pepsi are fierce competitors, they buy things like sugar, fruit and water from the same vendors. If it were found that a particular vendor is having significant product quality issues that could lead to health issues for consumers, it might help the entire industry if this knowledge was shared with minimum friction.

Dashboard, KPIs and Notifications

Ultimately, the objective of process automation and optimization provided through an ETM application is to centralize and streamline enterprise technology processes, such as technology lifecycle management, improving IT efficiency and productivity. To further this objective, the interface into an ETM application would need to be through a self-service portal that provides a configurable dashboard.

This dashboard would inform and measure KPIs that the CIO and perhaps other users like board members, CIO's staff, and heads of lines of business want to monitor. The dashboard would also capture notification of prioritized events. For instance, a secure offboarding process KPI might include the average time it takes to offboard employees, while a notification might provide an alert that a specific former employee hasn't yet returned a piece of technology like a laptop, or that there was an issue removing the employee from the payroll system.

Help CIOs Better Manage What They Do

Ultimately, an ETM application would help the CIO and IT better manage the ET processes that deliver the experiences and outcomes required by the business to succeed. This means an ETM application needs to help the CIO better manage enterprise technology by enabling more automated and optimized processes that utilize this technology, improving ET process efficiency, enhancing security and compliance, and delivering broader business observability.

| Improve ET Process Efficiency | Enhance Security & Compliance | Deliver Business Observability |

Figure 6-3: ETM applications help the CIO better manage IT

Improve ET process efficiency

An ETM application would help the CIO better manage the six things they do, improving overall ET process efficiencies. Here are few examples:

- **Manage for cost structure**: This is a big win for an ETM application, as process automation removes manual processes, while ETM notifications can inform the CIO when events occur that can impact costs.
- **Manage for service level**: KPIs could be built around quality of experience for both customers and employees to help ensure exceptional service levels continue to be provided by IT.
- **Manage for budget and schedule**: An accurate lens into the entire technology portfolio helps ensure upgrades and replacements are proactively purchased at lower costs than making emergency purchases.
- **Manage for continuous improvement**: KPIs are maintained that constantly monitor program status, providing alerts if workflows are veering off course.

- **Manage for speed**: By running a more efficient IT organization, money gets freed up to increase investment in potentially innovative, game-changing technologies.
- **Manage for disruption**: By making ET processes more efficient, budget gets freed up to invest in disruptive initiatives like digital transformation.
- **Manage for optimization**: By providing the CIO and staff end-to-end observability for their ET processes, they are much better positioned to identify optimization opportunities – perhaps assisted by AI/ML.

Enhance security and compliance

Before there is a security incident, an ETM application could provide metrics and insights into the current state of a company's security risk profile in an actionable way as defined by workflows that pull and aggregate data from point security management tools exposing security vulnerabilities within the technology portfolio. Has a terminated employee's access to an application that stores private customer data not yet been removed?

Then in response to an incident, an ETM application could run automated ET processes, inform responsible stakeholders of the impact of the breach, and communicate both externally and internally to help the company get services back up as rapidly as possible.

For companies that have a security platform that aggregates notifications and information from point security tools, this looks like another point tool to the CIO. With an enterprise technology management application, data and notifications from the security platform could be consolidated in an ETM dashboard so the CIO has holistic visibility into the entire technology landscape, including security. And if companies don't have a security platform, they could utilize the ETM application to consolidate data and alerts from the point security tools.

As a compliance example, with the trend to remote workforces, accurate tracking of all assets becomes even more challenging. Companies can't easily control employee use of company-issued devices, such as clicking on a phishing email sent to the employee's company email account. Installation and management of mobile device management (MDM) software at the time of asset purchase can help address this challenge, but as installing MDM software is typically a separate process from procurement, it often gets missed.

An ETM application could be used to automate the workflows so at device purchase, MDMs are installed so device data is tracked in a single source of record. During an audit, IT would know where to find each piece of technology.

And as part of a secure offboarding process, an ETM application workflow could trigger that company-issued devices be automatically locked with codes safely stored so the devices could only be unlocked when the device was received back by IT. Likewise, access controls could also automatically be removed for offboarded employees, reducing the risk they will do something nefarious like access customer information post-employment.

Rite Aid CIO, Justin Mennen, mentions that retailers like Rite Aid are heavily regulated and must meet compliance requirements such as SOX, PCI and HIPPA. As such, there are a lot of controls in place in his environment with well documented procedures. Some of these procedures are manual and some are automated.

Even with all these controls and procedures in place, Mennen mentions that there is a lack of available visualization across these workflows. For instance, Rite Aid has roughly 51,000 store associates, working in roughly 2,500 stores across the United States. Having enterprise-wide visibility into the status of onboarding and offboarding employees would be extremely valuable in ensuring these processes were done, and logged that they were done correctly.

Delivering enhanced IT and business observability

Taking a step back, Edward Wustenhoff provides his three steps to improve processes:

- **Observability**: First you must have observability. You can't manage what you can't see. And having end-to-end observability can help with troubleshooting.
- **Measure**. Second, define what you care about, which points to KPIs. Which metrics tell you whether you're doing a good or bad job? Put a number value on it and have an opinion about what is good and bad. For instance, on a scale of 1 to 10, 1 is bad and 10 is good. This implies you need to make the thing you want to observe measurable, so that goes to having access to the data.
- **Controls**: Third, create the ability to move the number value of your KPI in the desired direction. AI/ML can help here, using computers to provide insights to help you move the KPI in the direction you want.

That said, we all know that technology landscapes produce a tremendous amount of data that needs to be correlated and analyzed to derive insights that identify root causes for issues that can impact user experiences, data security risks, and barriers to the efficient execution of ET processes. By training AI/ML models with this data, an ETM application could provide real-time insights — a level of IT observability — that the CIO doesn't have access to today. These insights can help guide how the CIO manages the available controls.

Beyond providing enhanced IT observability, an ETM application would do well to take the next step of mapping ET process performance against business KPIs and metrics, providing improved real-time business level observability. An ETM application with this type of capability could clarify not only how IT is adding value to the business, but also help to proactively remediate issues rooted in broken technology or ET processes that run the risk of taking value away from the business.

For instance, Mennen shares that there are several capital allocation processes that are very visual across the whole of Rite Aid. There are many point tools available to help with spend analysis, such as breaking down how the different business entities manage OPEX and CAPEX spend. However, these tools are not well integrated. An ETM application could deliver this integration, providing higher level observability into things like company-wide spend.

KPIs would then be mapped to IT and business observability. Mennen says in his business there are operational KPIs more focused on managing day-to-day operations, and business KPIs that describe how the business is performing. An ETM application could help in reporting both sets of KPIs.

Among other things, as Anders Romare points out, the availability of KPIs through an ETM application dashboard would enable the CIO to drill down to justify technology investments.

In addition, as Mike Kelly observes, an ETM application that provided this level of observability to the CIO could also be used to provide radical transparency to the rest of the organization. "In my experience," Kelly shares, "this has long been a struggle for the CIO and letting consumers know what they spend on what services could provide a great foundation to build trust and credibility across the business."

Of course, the essential pivot in this discussion has been around the maturity of existing ET processes, ranging from undefined to manual to fully automated and self-optimizing. With this perspective, it's time we discuss a framework to help CIOs inventory the maturity of the ET processes they have running in their shops.

Chapter Summary

Key takeaways
ETM architecture includes:
- Connectors and data
- Automated and optimized processes
    - Built from software-defined workflows
    - Deploy real-time monitoring and alerting
    - Driven by AI/ML
    - Include connectors to third-party ETM applications
    - Built with open APIs
- KPIs and notifications
- Dashboard

The main benefits of an ETM application include:
- Improved ET process efficiency
- Enhanced security and compliance
- Deliver business observability

Key questions

- IT Observability: What are the KPIs and metrics you use to map enterprise technology process performance?
- Business Observability: What KPIs and metrics do you use to map enterprise technology process performance against business outcome objectives?

Chapter 7: ET Process Maturity Journey

At the end of the day, IT delivers services to the business implemented from processes executed through workflows. If the CIO is going to save costs by making ET processes that touch technology more efficient, then a natural question is, how mature are the existing processes used to deliver the IT services?

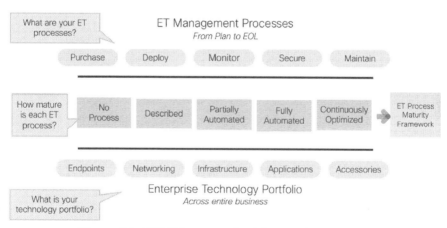

Figure 7-1: ETM Framework with essential questions

To answer this question, we return to the ETM Framework, as shown in Figure 7-1. We start by taking an inventory of your ET processes. After all, you'll need to know what your processes are before you can rate their maturity.

Each ET process can then operate at one of five maturity levels: No process, Described, Partially Automated, Fully Automated, and

Continuously Optimized. This provides the organizing principle around the ET Process Maturity Framework.

ET Process Maturity Framework

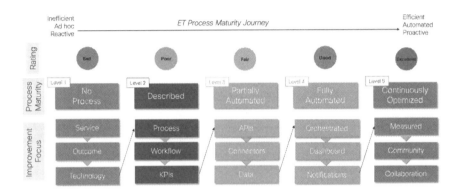

Figure 7-2: ET Process Maturity Framework

The ET Process Maturity Framework provides a way for the CIO to identify specific opportunities to improve ET processes, ultimately with the objective of delivering more value back to the business.

As shown in Figure 7-2, the ET Process Maturity Framework defines a five-level service maturity journey that can be applied to each ET process used to deliver IT services to the business. An IT organization can have several ET processes at all levels of maturity.

From ad hoc and reactive to automated and proactive

At the lowest level of maturity, the ET process is undefined and implemented in an ad hoc fashion, completely reactive to circumstances. For instance, if IT is asked where an asset is located during an audit and an IT person must play detective to manually find it without any defined process for guidance, this would be the lowest level of process maturity and likely produce the lowest level of IT productivity and efficiency.

At the other end of the spectrum, an ET process that is fully automated and proactively delivering value while continuously being optimized would be an example of a process operating at the highest level of maturity. A demand-forecasting service that deploys a process built from workflows that automatically update the status of all existing technology, and schedules POS device replacement purchases based on tracked failure rates might be an example of a highly mature ET process. A highly mature ET process is likely the most efficient, resulting in the highest level of IT productivity at minimum spend.

Let's explore each level of maturity in more detail.

Level 1: No Process

Description: No ET process is written down. The work is implemented completely ad hoc without a formal process described. Given there is no process, the maturity is at level 1.

Improvement Focus: Undefined-to-described
- **Service**: Formally defines the IT service to be delivered to the business entity
- **Outcome**: Formally describes the outcome this IT service should deliver. This will become crucial not only in describing the value of the service to the receiving business entity, but it will also help define if the service is delivering an appropriate level of value.

- **Technology**: Inventory the technology portfolio in scope required to implement the IT service. Ideally, a full inventory of the enterprise technology portfolio is done, creating a single source of truth. This is because it's best to start with an accurate inventory of technology that will be touched, as the full gamut of IT services will ultimately touch all technology.

Once these three actions are taken, the ET process moves to Level 2: Described.

Level 2: Described

Description: The ET process is described, but still implemented through manual tasks.

Improvement Focus: Described to partially automated
- **Process**: Describe the ET process used to deliver the IT service. Ultimately, the objective of the ET process will be to deliver the organizational goals as outlined in the IT service.
- **Workflow**: Define the workflow(s) that will execute the tasks required to implement the ET process. A workflow task can be automated and audited against compliance regulations. As a side note, if the workflow is described as part of a large PDF, then it will not be possible to move to Level 3: Partially Automated, as the workflow will need to be defined in software to be automated.
- **KPIs**: Defines the inputs, outputs and KPIs that will be used to measure how well the ET process is delivering on the organizational goals outlined in the IT service.

This more efficient approach could mean doing the work in Level 2, taking the ET process to maturity Level 3: Partially Automated.

Anders Romare shares that many IT organization ET processes he experienced are manual and not measured, so they don't really know how long these processes take to complete. A manual step might only take a few minutes to do, but the employee could get interrupted by a phone call and now those few minutes have expanded to a few hours of wall clock time. As another example, some processes might take a long time to complete simply because the employee lacked the proper training and didn't fully understand the work required.

As Romare points out, often in these situations the CIO and IT organization don't learn there are issues with the processes until the employees executing them start complaining that the workflow is not workable. There must be a more efficient way to get work done within the IT organization.

Level 3: Partially Automated

Description: The delivery of the ET process is implemented through a mixture of automated and manual tasks.

Improvement Focus: Partially automated to fully automated
- **APIs**: The APIs into technology touched by the service are identified.
- **Connectors**: Connectors into an ETM application that communicate with these APIs have been implemented.
- **Data**: The workflow that implements the process that delivers the service is written in software, sending and receiving data to/from the ETM application and the point management tools. At this step, this data is well defined.

If these three actions are taken, the ET process moves to maturity Level 4: Fully Automated.

Of course, automation isn't always the answer to higher IT productivity and lower costs. Melissa Gordon — current CAO at Tidal Basin Group — says she asks these basic questions when deciding whether to automate a process:

- How many hours does it currently take to complete this process?
- How much time is wasted waiting for data to transfer or information to be sent given a lack of integration?
- If we automated this process, how many hours and wall clock time would we save?

For instance, a process that currently takes 200 people to complete might save $400,000 a year if automated. Let's say you could hire a systems integrator to do the automation and integration of the process costing $100,000. The $300,000 savings could then be spent on other billable work, helping to fund Transformational or Performance Zone initiatives, or translate into not having to hire additional staff if you're growing, as the 200 employees are reassigned to different tasks.

The key, of course, is understanding the complexity of the process. If automating the enterprise technology process will be quick and save a lot of time and money, it's a different trade-off than having to kick off a one-year project.

This is where the availability of an ETM application comes in. Instead of having to hire system integrators to do custom integration, or allocating internal staff to do the same, you would utilize connectors and your own teams to build automated workflows.

Level 4: Fully Automated

Description: The workflow that delivers the ET process is fully automated. The ET process is holistically managed with complete observability into the relevant IT landscape of technology.

Improvement Focus: Fully automated to optimized
- **Orchestration**: Automated tasks are unified into an end-to-end ET process, allowing IT to manage the entire process lifecycle from a single location, including development, testing, monitoring, and measuring.
- **Dashboard**: Visualization is available through a dashboard where the measurement KPIs, SLAs and XLAs can be monitored.
- **Notifications**: Alerting is in place, automatically sending notifications to the dashboard and appropriate personnel upon re-established triggers.

If these three actions are taken, the ET Process Maturity moves to maturity Level 5: Continuously Optimized.

One advantage of a fully automated ET process is the ability to add failsafe conditions to the workflow. For instance, Melissa Gordon pointed out how the Krono's ransomware attack disrupted the delivery of payroll to thousands of employees.[16] The thing is, many employees can't live without their paycheck. And as paychecks are often delivered on a regular basis in regular amounts, a payroll workflow could be put in place with a condition something to the effect of: If the third-party payroll system goes down for any reason, automatically continue to deliver payroll using previous payroll amounts.

Level 5: Continuously Optimized

Description: Data is used to proactively deliver, optimize, and continuously improve ET process.

Improvement Focus: Optimized to community
- **Measured**: ET process is measured both in terms of delivery time and value creation to assist in continuously improving the process. AI/ML may be incorporated to help in this optimization function. For instance, a demand forecast process might identify patterns in purchasing that can be further streamlined.
- **Community**: As part of finding continuous improvement, a community of third-party resources are identified, such as IT outsources and non-competing IT organizations willing to share best practices.
- **Collaboration**: In this final step, continuous collaboration occurs within the community. For instance, if one member gets hit by a ransomware attack, this can be shared with the community to warn them of the attack and provide real-time information to improve all member defenses.

At Level 5, essentially the data becomes the new operating system (to paraphrase a visionary statement by Mike Kelly). Actions are completed because the data brings knowledge of what's happening in the environment, with perhaps a final manual approval if necessary.

Or perhaps you simply bypass the manual intervention altogether. For instance, if an EC2 instance hasn't been touched for, say, 90 days and there is no registered owner for the instance, an automated workflow might automatically shut the instance down. If someone complains, then the workflow could start the instance back up and now you can register the owner of the instance.

In general, people like shiny new things but are terrible at shutting things down. Does any IT company have a Shut Down Director? Where is the independent program management team and how good are they?

One technology company migrated infrastructure from its own data center to an Equinix data center and didn't move 34% of the applications, as it was discovered they weren't needed. That's 34% of applications that, for months and years, IT was using precious budget to pay for the applications' underlying compute, storage and network infrastructure, and conducting regular backups. Given IT is not very good at turning things off, having ET processes in place to do this automatically seems like a good idea.

As another example, a company implemented a quoting tool for the sales team and part of the tool included an elaborate approvals table. Unfortunately, when the tool went live, sales managers were being bogged down with approval requests. After reviewing the data, it turned out less than one half of one percent of quotes were not getting approved. So, a process was put in place to automatically approve quotes, taking the friction out of the workflow. Another process automatically works through deals that aren't aligned with the standard deal and sets a workflow to look at these deals more carefully. These are great examples of more mature, optimizing processes.

In terms of collaboration, to what extent would CIOs share best practices and knowledge with one another?

David Ching — former CIO at Safeway — is confident that CIOs would share experiences and insights. He adds that to some extent, CIOs already work together participating in different role-specific, industry and government sponsored groups.

Randall Spratt — former CIO/CTO at McKesson — shares a similar optimism around CIO collaboration, especially as it relates to security. He shares that organizations are increasingly "opening their kimonos a bit to talk to other companies, such as Microsoft, the federal government, or other companies that have deployed similar security technologies." When hit with a security incident, a CIO wants to know if they're first in line, if anyone else has seen this type of security threat before and what to expect.

ETM applications that communicate with each other can automate this sharing of crowdsourced knowledge and best practices for the ultimate good of helping out all CIOs.

So how mature are your ET processes?

Conduct an ET Process Maturity Assessment

An IT organization of any size likely has hundreds of ET processes running at different levels of maturity, with some processes being more important than others.

For instance, Alain Brouhard shared that in terms of employee offboarding, Coca-Cola had a fully defined process. However, this process was more of a manual to-do list of tasks to be completed. There was no automation. There was no process optimization occurring. With more than 10,000 employees, it became an interesting question of cost savings. How much would Coca-Cola save by improving the maturity of its offboarding process alone?

So how does a CIO prioritize investigating ET processes that may be operating sub-optimally?

Value pillars provide a logical on-ramp.

Value pillar on-ramps

Figure 7-3: Value pillar on-ramps

Of course, a company wouldn't start by assessing the maturity of all ET processes. That would be too much out the gate.

All CIOs, to some extent, prioritize their attention based on current pain points. Do they need to figure out a way to fund digital transformation? Did they recently have a security breach? Are they struggling passing audits? Is the CIO worried they are not meeting compliance regulations given a lack of data hygiene?

As shown in Figure 7-3, there are five universal IT value pillars where most IT organizations can focus their attention to bring value to the business, which map to potential pain points experienced by the CIO:

- **Financial**: This is where the CIO can help the business better utilize available spend, for instance in providing more accurate demand and refresh forecasting. This is also where the CIO can free up budget to find funding for Transformation Zone initiatives such as digital transformation.
- **Experiences**: This is where IT helps deliver exceptional customer and employee experiences, critical to the success of any modern organization.
- **Security**: Given the continuing rise of cyberattacks, keeping the company's data safe is a top priority for every CIO.
- **Compliance**: This is where the CIO ensures the company is complying with relevant regulatory requirements such as GDPR, CTPA, HIPAA and ISO 2001.
- **Audit**: Related to compliance (but typically delivered by a different part of the organization), this is where the CIO ensures the company is ready for an audit.

Start at a value pillar

A CIO would do well to start by exploring one of the five value pillars to see where their organization might gain the most value from improving the maturity of their ET processes. Over time, CIOs of the most mature organizations would have explored all five.

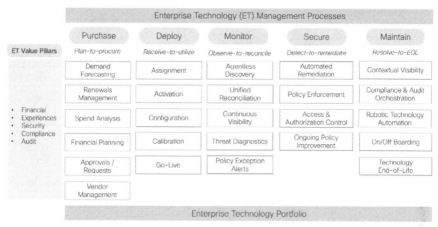

Enterprise Technology (ET) Management Processes				
Purchase	Deploy	Monitor	Secure	Maintain
Plan-to-procure	*Receive-to-utilize*	*Observe-to-reconcile*	*Detect-to-remediate*	*Resolve-to-EOL*
Demand Forecasting	Assignment	Agentless Discovery	Automated Remediation	Contextual Visibility
Renewals Management	Activation	Unified Reconciliation	Policy Enforcement	Compliance & Audit Orchestration
Spend Analysis	Configuration	Continuous Visibility	Access & Authorization Control	Robotic Technology Automation
Financial Planning	Calibration	Threat Diagnostics	Ongoing Policy Improvement	On/Off Boarding
Approvals / Requests	Go-Live	Policy Exception Alerts		Technology End-of-Life
Vendor Management				

ET Value Pillars

- Financial
- Experiences
- Security
- Compliance
- Audit

Enterprise Technology Portfolio

Figure 7-4: Example ET processes

In Figure 7-4 we highlight ET processes common to most businesses, organized within the ETM Framework. Any of these processes can touch one or more technologies, but one or more might be further places to start your discovery.

In addition, often an ET process will cross value pillars. For instance, in the secure employee offboarding example, a process might help reduce financial burden by automating a terminated employee's access to various applications. In addition, the automation of the offboarding process helps with compliance by ensuring the terminated employee no longer has access to customer PII data available through, say, the marketing automation platform. And by using software to implement the process, the ETM application workflows can be logged, which can be very helpful during an audit.

Leverage an ETM application to help with ET Process Maturity assessment

Interestingly, an ETM application could be the perfect application to help with an initial ET Process Maturity assessment by leveraging its design and documentation capabilities. Let's say you wanted to first assess the maturity of the secure onboarding and offboarding processes and discover that they are at Level 1: No Process. You could use the ETM application's drag and drop interface to describe the processes workflow — in software — taking them to Level 2 of the ET Process Maturity Framework.

To then get to Level 3, you could effectively turn on the ETM application's automation capabilities to automate the workflow, at least partially. This would include setting up the connectors to connect the ETM application to the relevant point management tools. You would also use the ETM application dashboard to provide visualization into the KPIs and established notifications.

Once you've seen the value of doing this for onboarding and offboarding employees, you could then start prioritizing other processes of interest such as privacy management, security, or demand forecasting.

The CIO's challenge

Here's the CIO's challenge: Take 65% of cost out of your Productivity Zone initiatives by improving the maturity of your ET processes, potentially freeing up millions of dollars of budget to refocus on disruptive technologies like digital transformation.

Chapter Summary

Key takeaways
- Leverage the ET Process Maturity Framework to identify the maturity of your ET processes

- Conduct an ET process maturity assessment to identify ET processes that are leading candidates for automation.
- Identify the value pillar that makes most sense for your organization to focus on first with this ET process maturity assessment.

Key questions
- What are your ET processes?
- How mature are these processes?
- Do you accurately know what's in your technology portfolio?
- In which value pillar (finance, experience, security, compliance, or audit) are you experiencing the most pain?

Chapter 8: CIO Do's and Don'ts

If an ETM application existed, this could help motivate a group of Do's and Don'ts for a CIO to consider. Here are a few:

Stop thinking CMDB is the answer to technology management

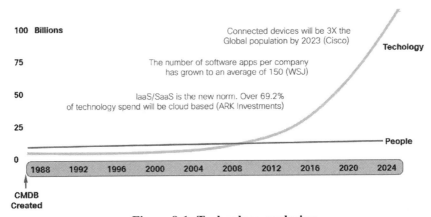

Figure 8-1: Technology explosion

Explosion in technology

We all know there is an explosion in technology used by businesses, and the trend is only for more technology, not less. According to Cisco, connected devices will be 3x the global population by 2023.[17]

Think about all the Apple watches, Whoop bands and other health-related devices on the market. Think about all the retail experiences that people are trying to provide that include in-store wireless access points measuring traffic and software measuring loyalty. Think about all the remote workers essentially running their own "in-house" data centers with endpoints, servers, storage, networking, and applications at risk of a cyberattack. Now add developers and contractors readily bringing up virtual instances and makeshift environments to create, test, and augment digital business initiatives. Let's not forget about companies creating new or refactoring legacy on-premises applications into cloud native applications built on distributed microservices, open-source apps and externally connected services. It's a cascade of technology and it's happened quickly, with most of the explosion having occurred relatively recently, as depicted in Figure 8-1.

IT is left with the responsibility of managing and securing all this technology, with the objective of maximizing value received throughout its usefulness to the business. Enabling this work requires some form of database to keep track of all the technology: a highly accurate single source of truth for the data that describes a company's entire technology portfolio. Afterall, you can only manage the technology that you can see, as Peter Drucker reminds us: "If you can't measure it, you can't manage it."

CMDB is purpose-built for service management

For the past three plus decades, this database has been the Configuration Management Database (CMDB) — first introduced in the 1980s as part of ITIL 2. A CMDB is a file — usually, in the form of a standardized database — that contains all relevant information about the hardware and software components used in an organization's IT services and the relationships between those components. A CMDB provides an organized view of configuration data and a means of examining that data from any desired perspective.[18] This configuration data can include relationships and interdependencies between items, the history of changes to each item, and attributes for each item (such as type, owner, and importance).[19]

The CMDB was originally conceived during a time when an organization's technology landscape was significantly less complex than it is today. For instance, back in the 80s and 90s and before cloud computing, IT might provide an email service, consisting of a cluster of servers running Microsoft Exchange and deployed in the company's own data center. To help with servicing, the IT organization needed to keep track of which servers were running the email service, which were attached to which databases, and which were stored on which hard drives and so on. This is because if something went down anywhere in that chain, the email service could be impacted.

In other words, to assist the IT organization in delivering the email service, it greatly helped to understand the dependency mapping between different assets. This was a primary motivation for building the CMDB. To help with this mapping, the CMDB has configuration items (CIs). As defined by ITIL 4, CIs are "any component that needs to be managed in order to deliver an IT service." As such, every CI has mapped relationships to each other.

In addition, back in the 80s and 90s, the dependency mapping was kept relatively constant. Once the email service was up and running, the IT organization didn't really touch it, following the proverbial IT Rule #1: "Never change a running system." Or put more colloquially: "If it ain't broke, don't fix it."

The same IT organization might also run an ERP system, which had its own set of dependencies. But back then, the ERP system and email service were self-contained and kept separate, likely running on different sets of servers. So again, the dependency mapping for the ERP system was kept relatively constant.

As a result, once the CMDB was designed to hold the information for the email service and ERP system, including potentially utilizing a professional service to extend the CMDB to meet requirements unique to the organization, there would be little reason to touch this configuration.

And when a change to a running service was needed, such as installing a patch or upgrading an operating system, the IT organization would go through a proper change and release process. One fundamental question that needed to be answered was: If a change was made to a running system, then what were the dependencies? If we reboot a server, for instance, will this create an outage, or is there enough redundancy built in so we could reboot server one, then server two, then server three and so on? This information and these dependencies are all kept in the CMDB to assist the IT organization in delivering services back to the business.

The CMDB is not purpose-built to be a general data warehouse for enterprise technology in the modern era

Today, however, not only are technology landscapes more varied, but they are also more dynamic.

For instance, migrating to the cloud continues with Gartner predicting, "More than half of enterprise IT spending in key market segments will shift to the cloud by 2025."[20] And with cloud computing comes flexibility to bring up or move workloads with relative ease. A virtual machine running a company's ecommerce system could sit on a company's on-premises server one moment, but then more virtual machines running the ecommerce system could be spun up on a public cloud to handle additional customer demand, as is often the case during a holiday shopping season.

Likewise, the systems an IT organization utilizes to deliver services are no longer kept isolated, making the dependency mapping much more complex and dynamic. A virtual machine in one moment could be running multiple IT services; therefore, a change made to that virtual machine, such as installing a security patch, could impact multiple IT services. But then in the next moment, a service could be moved to a different virtual machine running on a different cloud provider's infrastructure. Likewise, SaaS, the largest public cloud adoption by enterprises[21], also introduces external dependencies and costs that need to be managed under the CIO's purview.

Consequently, an IT organization's dependency mapping of its entire technology portfolio could end up looking more like an unorganized wiring closet full of intertwined network cables, but with port connections that seem to frequently change or disappear into thin air.

This dynamic and complex technology landscape means data about CIs is constantly changing. Consequently, the initial CMDB designs may need to be revised regularly, making frequent, unplanned maintenance of these relationships a constant undertaking by the IT organization. If the CMDB is not kept up to date, then it becomes harder for the CIO to trust that the CMDB is accurately reflecting the organization's entire technology landscape. This can make it difficult for CIOs to view their CMDB as a purpose-built data warehouse for their entire technology portfolio.

Of course, a CIO can elect to continue to spend money on professional services to continue to update and extend their CMDB database in an on-going attempt at keeping the CMDB aligned with the organization's actual technology portfolio. But this is analogous to painting the Golden Gate Bridge: By the time you're done painting the bridge on one end, you have to start all over again painting the other end.

Another challenge CIOs face in trying to view their CMDB investment as a purpose-built data warehouse is the need for integrations with all the point management tools. Without these integrations, the CMDB can't be automatically and seamlessly fed the data that it needs to stay up to date. Keep in mind, merely sending point management tool-data to the CMDB does not address the correlation needed for such items to avoid conflicts, duplications and inaccuracies. In many cases, the IT organization must figure out these integrations on their own, which results in the CMDB not being uniformly connected to a company's entire technology portfolio.

Rather, what the CIO needs is a purpose-built data warehouse for its entire technology portfolio. This is an attribute of an ETM application, running a database that is purpose-built from the ground up to directly integrate to point management tools and other sources to handle today's dynamic and varied technology landscape. This is not to say an ETM application couldn't be built on a CMDB; but rather to observe that it is likely easier to build an ETM application on a database purpose-built to meet the ETM application requirements.

This is also certainly not to say that IT organizations should throw away their CMDB investments — not at all. Support of IT service management (ITSM) applications appears to be a natural use case where CMDB investments can thrive.

The complementary nature of ITSM, CMDB and ETM

ITSM has been described as the activities that are performed by an organization to design, build, deliver, operate, and control information technology services offered to customers.[22] Put another way, ITSM describes how IT teams manage the end-to-end delivery of IT services.

ITSM grew up in the 1980s, about the same time as CMDB. CMDB and ITSM are closely linked. One goal of a CMDB is to provide an organization with the information needed to make better business decisions and run efficient ITSM processes.[23]

In fact, ETM and ITSM share the complementary core belief that IT should be delivered as a service.

However, architecturally, a fundamental difference between ITSM and ETM applications is typically in the database. ITSM applications are likely built on a CMDB database. So even though IT Asset Management (ITAM)[24], as an example, is often included in an ITSM application portfolio, the CMDB can restrict the ease at which an organization can track new technologies incorporated by the business without having to bring in expensive managed services or custom integrations to update the CMDB.

On the other hand, an ETM application would likely be built on a purpose-built data warehouse designed from the ground up to support the broad range of technologies in use by modern businesses. Such an ETM application would provide additional benefits, such as:

- Allowing an organization to continue to use existing point tools known, liked and managed by IT operators and applied across different domains and technologies
- Enabling the organization to aggregate, correlate and normalize operating data sourced from the many point tools to support broader technology management use cases and to automate a broad range of business processes.

This means a vendor providing an ETM application would also have to provide the integrations needed by the IT organization to connect with the technology landscape. This would be a critical capability; otherwise, the IT organization would be back in the business of having to stitch solutions together.

This also implies that the mapping between the data fields in the point tools and the ETM application database would likely be done through the ETM application's user interface and dropdowns. The ETM application would have to be built to not require engaging professional services to make custom changes to the ETM database.

From an ETM perspective, ITSM represents a set of important management tools and applications that operate predominately with processes described in the Maintain section of the ETM Framework. In fact, by one estimate, ITSM is mostly deployed in support of customer experience (35%) and service quality (48%).[25]

On the other hand, an ETM application would be designed to support technology management of ET processes across the full lifecycle of processes — including Purchase, Deploy, Monitor, Secure and Maintain — that touch the entire enterprise technology portfolio from planning to EOL/retirement/disposal of assets. From a technology stack perspective, an ETM application would likely run on top of the

ITSM application, along with the variety of other tools and applications deployed by the organization to manage technology silos.

One way of thinking about the difference is ITSM is typically associated with service desks and human workflow. On the other hand, an ETM application would likely be more associated with automated technology workflows.

For example, a typical ITSM use case involves submitting tickets with relevant information, such as in response to an HR request for a laptop for a new employee as part of an onboarding process. This ticket submission kicks off a set of activities, which lands the ticket in an IT team's queue where incoming requests are sorted and addressed according to importance.

That said, an ETM application is not trying to replace a company's ticketing system, but rather augment it. In fact, it would likely make a lot of sense for there to be an integration between the ITSM and ETM applications.

Following this onboarding example, it might make sense for the onboarding workflow to be described and implemented in the ETM application. The ETM application could send the ticket request for a new laptop to the ITSM application through an automated workflow.

But the ETM application could also interact with other tools and applications, such as installing an MDM on that new laptop, establishing a VPN service for secure work from home, setting up a login so the employee can participate in web conferencing on day zero, and providing access to other SaaS applications and on-premises resources necessary to be a productive employee.

Similarly, the ITSM application could provide valuable data back to the ETM application, such as the ticket status (e.g., a ticket was completed or not completed after a set amount of time).

New employee onboarding is but one of many possible use cases. For example, the same ITSM and ETM application relationship can be applied to the more complex offboarding processes which can require tasks such as asset access termination, data reassignment, endpoint and other device reclamation, data retention, sanitization, reallocation, and retirement. An ETM application integrated with ITSM could not only streamline processes, but also negate the gaps and omissions that can present CIOs with significant financial, security and compliance risks.

In a similar fashion, the ETM application could provide end-to-end orchestration in support of the ITSM application. For example, if the ITSM application needs data from a tool used to manage company laptops, the ETM application could act as a conduit for this data through integrations with the ITSM application, the laptop inventory management tool, or even a third-party reseller that provides laptops and other predetermined new hire kits. This would keep the IT organization from having to stitch custom integrations together to enable that process and would ensure an audit trail of such service delivery.

ETM complements TBM

While we're talking about complementary technologies and frameworks to ETM, it's worth noting that the Technology Business Management (TBM) council will show a slide positioning TBM as "the technology exec's application." Formally, TBM is defined as a discipline that improves business outcomes by giving organizations a consistent way to translate technology investments to business value.[26] TBM provides a framework and methodology focused on cost transparency, identifying the total cost of IT, and shaping demand for IT capabilities. As such, TBM provides a standard cost model taxonomy.

Founded in 2012, TBM has developed a taxonomy for how IT costs can be explained back to the business. TBM's genesis was when the prevailing view of IT was "you guys take too long, and you cost too much." As Mike Kelly shares, CIOs back then had to sit in the hot seat at the C-suite meetings and defend their turf. As a non-profit consortium, BPM gave CIOs industry best practices to communicate the cost, quality, and value of IT investments to their business partners. TBM was designed to help IT leaders gain deeper insights into IT spending, budgets, and resources.

According to Tim Pietro, FinOps & Technology Business Management (TBM) Practice Head at Capgemini Invent, TBM assists in organizing IT assets, such as infrastructure, documents, people — everything we do in IT — into a hierarchy to enable conversations around services, capabilities and where a company should invest its IT budget.

On the other hand, according to Pietro, ETM is highly complementary, providing a framework to assist in operating IT.

For instance, through a connector between an ETM application and a TBM application like Apptio, data could be exchanged to improve the information sharing across the organization. As an example, given the ETM application holds an accurate, single source of truth on all assets, this data would likely help make Apptio's IT spend calculations be more accurate, such as sharing license usage information. Likewise, Apptio budget information would likely be helpful inputs to an ET process that's automating the advanced purchasing of POS devices due to anticipated failure rates.

Deliver IT-as-a-Service

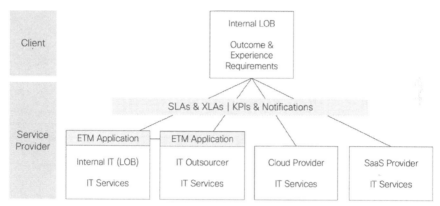

Figure 8-2: Internal IT as an LOB

Increasingly, a company's lines of business (LOB) — such as sales, marketing, finance, operations, and human resources — have a choice of IT service providers, whether that's using their own internal IT organization or outsourcing to an external IT outsourcing company, SaaS provider or cloud provider (see Figure 8-2).

To compete in this environment, the internal IT organization would do well to think of itself as a line of business, delivering IT-as-a-service (ITaaS) to the rest of the company. The transformation of an internal IT organization from operating as a cost center to a services-based operating model would likely produce improved levels of business agility for the enterprise, while enabling the CIO to focus efforts on the continuous improvement of the delivered services.

With an ITaaS approach, the internal IT organization would likely place greater emphasis on delivering the outcome and experience services required from the buying LOB, given the competition against external providers. This would help drive the need to measure value delivered through these services, assisting the CIO in articulating value delivered to the business units. And given this competition, IT would also place greater emphasis on continuous improvement, reducing its costs through improved operating efficiency and employee productivity.

Services value delivered through operating processes

However, before an IT organization can transform itself to delivering ITaaS, they must first define service. As former Cisco CIO Guillermo Diaz counsels, "To transform to a services-based operating model first requires defining the taxonomy around, what is a service?" This can be quite challenging as the term "service" can be used rather loosely by an organization to include non-service components like scripts and applications.

Diaz describes what an IT service is through a simple example:
- IT delivers an operational process to the business, such as *quote-to-cash (QTC)*
- The quoting component of that process is a service: a quoting service.

This implies a services-based IT organization would do well to also optimize around its operating processes. After all, a line of business will no doubt benefit from having a quoting service, but what they really need is for the quoting service to connect with an end-to-end process that completes with cash in the bank.

Continuing with this example, at a high level, a quote-to-cash process can include many systems, platforms and applications, such as:

- **Customer relationship management (CRM)** to keep track of customer relationships and ensure all the relevant data is available to support signing contracts, billing, collecting, as well as the post-sales analysis
- **Configure Price Quote (CPQ)** to enable the creation and quoting of customized design products
- **Enterprise Resource Planning (ERP)** to maintain product inventory

So, although the quoting service might be delivered through the CRM system, a robust, end-to-end *quote-to-cash* process delivered by the IT organization will likely be deployed through a combination of services, platforms and applications.

Here is where an ETM application can help. For example, if integrations exist with the ETM application and the various components that together deliver the QTC process, then higher level monitoring and measurements can be easily applied to that process, such as measuring cycle time and sending notifications around potential bottlenecks.

And if ETM applications were connected between the IT organization and an external service provider who is delivering a particular service — such managing an outsourced CRM system — then this higher-level monitoring and measuring could still be provided through the IT organization's ETM application.

This ability to share metrics and notifications would go a long way toward delivering radical transparency, as Mike Kelly observes.

Cisco's transformation to delivering of ITaaS

According to Guillermo Diaz, former CIO at Cisco, back in the mid-2000s, Cisco ran what they called a 'client funded' model where every major line of business within Cisco effectively had its own IT budget. This approach provided each LOB with its own funded IT resources, which in theory provided IT support more aligned with the unique needs of each business unit. Organizationally, this approach also drove Cisco's IT organization to silo itself into groups devoted to each line of business, with all groups leveraging a common infrastructure team.

What Cisco IT discovered was these organizational silos created several operating inefficiencies. For instance, in total, Cisco IT was running over forty instances of Salesforce, as each IT team delivered its own CRM instance to support the various business units. In addition, customers were experiencing many Ciscos, not just one, often based around how Cisco was structured. For example, Cisco customers would receive several different emails from the various businesses — all with a different look and feel.

Guillermo and his team took a step back and said, "Let's really look at this; we can do better."

They decided to implement a services-based operating model within IT, delivering IT-as-a-Service back to the business units. With this approach, Cisco IT would organize itself to act more like a service provider instead of mandating what the business units needed to do, or just taking orders.

By transforming to a services-based approach, Cisco IT could more easily improve operating efficiencies by focusing attention on the continuous improvement of these delivered services, such as accelerating time to value and driving down costs. Cisco IT could also more easily measure the value delivered by these services, providing transparency into outcomes and experiences delivered by these services.

To help make this happen, Cisco IT developed a framework they called BOST:

- **B**usiness vision and policy: What is it?
- **O**perating processes: How do you operationalize the business vision and policy?
- **S**ystems (not the technologies per se), such as CRM or ERP utilized to deliver the service
- **T**echnologies that support the systems, such as the network and other key infrastructure capabilities

With this framework in hand, Guillermo and his team started asking questions like: "What are the services that we deliver? And what are the operating processes, systems and technologies needed to implement those services?"

When Cisco IT first looked at their entire delivered services portfolio through this lens, they discovered that they were operating literally hundreds of so-called "services" and multiple instances of the same service. For instance, when reviewing the delivery of a customer sales service, Cisco IT had to ask: Do we really need over forty instances of Salesforce? The answer, of course, was no. With this answer in hand, Cisco IT then reduced the number of Salesforce instances, saving Cisco millions of dollars, while not impacting the customer sales service delivered back to the business units.

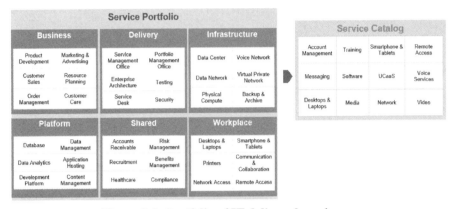

Figure 8-3: Portfolio of IT delivered services

Starting with hundreds of delivered services, Cisco IT settled upon a portfolio of 36 fundamental services that IT delivered, which have been adopted by TBM, as shown in Figure 8-3. This portfolio was then published in a services catalog that facilitates the business units ordering these services.

Be a Catalyst for Business Value

IT (almost by definition) finds itself in a position devoting a lot of human capital to things that people don't see as adding much business value to the organization; the work is seen more as an administrative necessity.

As Diane Randolph puts it: "Keeping accurate records of the technology used by the business should be easy, right? Yeah, the devil's in the details. Consequently, people outside of IT have difficulty understanding why managing technology is complex and why it is a problem — what IT's value-add is."

This lack of understanding perpetuates the view of IT as just a cost center (i.e., those teams over there that look after technology), when truthfully, many other business units within the company also aren't in the business of generating revenue and are also cost centers from a P&L perspective.

On the other hand, every external IT outsourcer playbook invariably includes being strategic advisors to their customers. They thrive on being technology experts to help customers deliver outcomes at reduced costs.

Likewise, IT needs to also think of itself as a strategic advisor to its peer lines of business, strengthening business-to-IT relationships with eyes on delivering business value.

After all, digital transformation creates change in both technology and how the business is run. For instance, according to a Gartner survey, nearly three-quarters of technology purchases are funded, at least in part, by business units outside of IT.[27] Generally, CIOs are not opposed to the business leaders taking more ownership of technology spending, but support needs to come from IT as the technology experts within the organization. This creates a natural incentive for business units to look to internal IT for expertise.

In fact, it's in everyone's interest that IT has a strong, advisory role with the other business entities. IT can help advise on the technology purchases, while also having input on technology spend that IT will likely end up supporting. IT taking more of an advisory role will become especially important as few business leaders are experts in security, which is becoming a critical consideration when bringing new technology into the business.

When there isn't support or guidance from IT, technology spend by the other lines of business can become fragmented through multiple

SaaS vendors across business units and geographies. This makes managing the spend to optimize value received challenging. For instance, there is risk of duplicate spending by different business units. When IT is brought in cold after technology purchases, all the rogue technology purchases can become a real mess to figure out and likely not result in the most efficient use of IT resources.

With an ETM application in place, IT maintains a precise record of all technology in use by the company. IT can make sure to not only avoid duplicate spending, but also provide data on the reliability of different vendor offers. IT can put workflows in place so when the technology comes in, it's immediately integrated into internal processes like providing access controls and monitoring license usage. The workflows could ensure data governance policies are being followed. Through the ETM application, IT could also monitor the value received from the technology spend against plan, which among other things can help guide further spend. And because IT can't own all business processes, it can set up the business owners with automated processes to improve overall efficiencies.

Of course, as Mike Kelly shares: "For IT to be seen as a value catalyst for the business, it can't be seen as delaying business units [from] moving forward with technology purchases they need to be successful." Going to IT for advice can't feel like going to the DMV for a new registration — slow and unresponsive. The role of IT is not to tell the business entity they're doing the technology all wrong.

"Shadow IT, for instance, can be very effective if the innovation is happening as close to the business problem as possible and people are coming up with solutions — that's great," continues Kelly. "Here the role of IT is to make sure the technology solution can grow and scale appropriately, as well as be compliant, not exposing the company to unnecessary risk."

IT can help get optics around the solution to see if there are better ways to solve the problem without slowing down the initiative by introducing bureaucracy, which doesn't solve anything.

And with the help of an ETM application, IT can be extremely agile with requests about technology using precise information about the company's technology portfolio available with a few mouse clicks. IT can discuss how the technology will be integrated into existing ET or set up new processes.

In other words, when asked what the IT strategy is, the answer should be, we don't have an IT strategy; we have a company strategy. Initiatives hang off those business strategies that cascade down into projects. IT is a trusted advisor around technology, whether the initiative is operated from within IT or a business entity. In addition to providing Productivity Zone services, IT delivers services that add value to these Performance and Transformation Zone projects.

Gamify IT

According to Red Hat CIO, Mike Kelly, one objective of the CIO is to help make employees EPIC in every function — effective, productive, innovative, and collaborative. After all, what CIO doesn't strive for some form of operational excellence within their organization?

And let's be honest, people who work in IT are cool. One way to recognize this and to help make them even more EPIC is by rewarding their EPIC behavior.

Rewarding employees for jobs well done inspires them to work harder and be more productive. Data from the Harvard Business Review suggests that 82% of Americans don't feel their supervisors recognize them enough[28] for their work. Additionally, 40% of Americans also stated they would put more effort into their work if they were recognized more often.

Employees in Amazon fulfillment centers get SWAG bucks for meeting certain goals which they can spend on actual products. For years in manufacturing, employees that come up with ideas to improve productivity get rewarded. At Red Hat, IT employees can win what they call a Plus One award for demonstration of operational excellence.

That said, it can be challenging to intrinsically motivate employees to put in more effort. Ideally, this motivation comes from enjoyment and job appreciation, not from cash alone.

To address this challenge, a comprehensive reward and recognition program that increases employee productivity would likely have these characteristics:[29]

- **Employees know how to earn rewards**: Ambiguous rewards guidelines can frustrate and alienate employees.
- **Tie reward to performance standard**: Employees must understand how their productivity and behaviors influence business outcomes. Employee rewards can demonstrate this on a more personal level.
- **Offer unique rewards**: Every workplace has its own unique culture, so the rewards should reflect items of value without your organizational setting.

- **Recognize large and small achievements**: Too often, only major accomplishments are rewarded, while smaller milestones achieved along the way (to making the big win happen) are often ignored.
- **Incentivize teamwork**: A great rewards program should focus on engaging and rewarding team efforts. It goes without saying that teamwork can amplify individual productivity, while strengthened relationships within and between departments and business entities can improve overall collaboration, innovation, and communication.
- **Gamify rewards**: Competition between peers can be unhealthy; however, playing the game of improving productivity where there can be multiple winners can help further drive teamwork.

A major way to improve IT productivity is through automating and optimizing ET processes. Therefore, deploy a reward and recognition program that rewards employees for coming up with ideas on ET processes that remove manual labor.

After all, as David Ching shares, "The people on the floor probably know a lot more about how to improve things than their management layer that, in a way, covers up this knowledge." Employees working in the trenches implementing manual processes can be fantastic sources of inspiration for the creation of ET processes that would make them more productive.

In fact, Ching says some of his leadership secret sauce came from walking around talking to employees in the IT organization to gain their perspectives. But the modern era of remote workers makes walking around and randomly peeking into cubicles less accessible.

An ETM application could help with this discovery. Without ETM, employees likely couldn't imagine the possibility of automating processes that touch different technology and organizational groups. With ETM's simple drag-and-drop interface, employees could conceivably create their own productivity-improving workflows, likely working with the other employees involved with the identified process to improve.

The owners of these workflows would be captured and every employee that creates a workflow that improves productivity by a certain amount can be rewarded.

Build Teams Aligned with Automated ET Processes

Traditionally, IT and its supported lines of business had more of a 'throw it over the wall' working relationship. The LOB hands over a set of high-level requirements and IT goes off working behind the curtain to figure out how to build a solution that meets these requirements. Of course, there are a lot of inefficiencies built into this approach, such as running Performance Zone programs.

Johnson & Johnson CIO, Jim Swanson, shares he and his team have implemented a better approach. They combine building a high performing matrix team with a hire-to-retire training mentality and an out of the gate workflow perspective.

For example, when building a tech team you might include a network engineer, full-stack developer, user experience designer, and a scrum master. On the LOB side, you might have the commercial leader, R&D scientist, and procurement professional.

Swanson shares how they train this team from a hire-to-retire perspective, which helps create even higher levels of performance. After all, nobody completes training if they don't see value in it.

However, when training is aligned with both individual career plans and the specific requirements needed for the team to be successful in delivering on its program, team members are more motivated to produce more for the company. This is especially true when you add incentive planning into compensation and promotion opportunities.

That said, Swanson reminds us that "you still need to connect the dots on how technology will work into the program, otherwise you won't end up with good results."

This leads us to the enterprise technology processes tied to the program and team. Ideally, the team cannot only define the ET processes used by the program, but also modify the workflows as they run the program in production, learning along the way. The successful running of these workflows provides an important benchmark where the rubber hits the road, and the team composed of members from both IT and corporate functions measurably demonstrate aligned outcomes and successful value creation to the business.

Given the technology used by the program in play could span the enterprise technology portfolio, the building of these process workflows must be quick and easy to do for the team members. This is where an ETM application would come in, using a simple drag-and-drop interface to not only build but also modify workflows over time. This would be a better solution than, for instance, bringing in a consultant to modify a CMDB to accommodate any new technologies required by the program.

Swanson also points out (from his perspective working in the pharmaceutical industry) that there are different layers in the technology stack to consider:

- Data science used to build and apply model outcomes in key decisions made across the enterprise
- Technology utilized by the enterprise
- Data produced by the technology and utilized by the data science

All three layers must work together to extract the most value from the technology purchases. If one level is in bad shape, or if the different levels aren't connected and communicating with each other, then decision-making can't be guided by informed data science.

Again, an ETM application can be helpful, for instance, by using ET processes built by the cross-organizational teams to automate pulling data from siloed technology, then enabling the data-science layer to access this data through the ETM APIs or connectors. Without an ETM application, the team would likely have to stitch a custom solution together, which not only adds cost and causes time delays, but can be a nightmare to support and update as the technology evolves.

Chapter Summary

Key takeaways
- Stop thinking CMDB is the answer to technology management
- Transform IT to a services-based operating model
- Be a catalyst for business value
- Gamify IT
- Build teams aligned with automated ET processes

Key questions
- How are you doing technology management today?
- Are you viewed as a trusted advisor to the rest of the business?
- How are you articulating IT's value to the overall business?

- How are you motivating your organization to improve productivity?
- How are you organizing teams that build, maintain, and optimize enterprise technology processes?

Chapter 9: Empowering Autonomous IT

So, where is this headed? If the industry were to create an ETM application, how would we define success?

You don't have to look very far to see another industry with the objective of removing manual operation in favor of automation. Think: autonomous, self-driving cars. Sure, they are set up to have manual intervention if needed, but they would likely define success as a time when all cars on the road are 100% self-driven — everyone just kicking back and enjoying the view.

As another example, we already have lights-out manufacturing. Spend a day at a Chrysler factory in Detroit and you'll find two million-square feet of robots automatically doing the work of building trucks. And the factory is as clean as a hospital.

Amazon Smart Warehouses

Amazon smart warehouses (aka fulfillment centers) offer another example. They start with their deep learning AI to predict what people are going to buy. The algorithm makes assumptions about you based on your age, location, socio-economic background, and purchase history. Then before you click "add to cart", the algorithm anticipates what you'll want in the near future and stocks your local warehouse with these products you're likely to buy. For instance, in 2020 the algorithm correctly predicted a high demand for face masks. We all know what happened there.

Once you've clicked buy, the autonomous, self-driving begins. It used to be that Amazon workers walked ten-plus miles per shift manually picking up boxes from shelves. This is no longer the case. Amazon's modern fulfillment centers are now controlled by robots that move entire shelves of products (known as pods) to the human pickers. The robots don't run into each other as they're controlled by a loosely described AI-driven air traffic control network that coordinates the route for every robot.

This technology arrived because of Amazon's acquisition of robotics company Kiva Systems — a leader in warehouse automation — for $775 million. It has been estimated that Amazon warehouses can now hold 50% more stock and retrieve that stock three times faster, reducing the cost of fulfillment by 40%. I suspect Amazon will define success when its fulfillment centers are 100% automated.

Reduce Productivity Zone costs

IT organizations are faced with a similar challenge that Amazon faced when fulfillment processes were largely done manually (probably operating at Level 2: Described on the ET Process Maturity Framework). Many of IT's ET processes, especially those operating in the Productivity Zone, are likely at Level 1 or 2 — either not defined or described in large PDF documents. How much spend could an IT organization save if it moved all its ET processes to Level 3, 4 or 5 – toward being an autonomous, self-driving IT organization?

To be clear, we're talking about only automating processes that can benefit from being automated. It may still be more cost effective to have a human install a new blade in an on-premises rack versus implementing a picker system that does it automatically. Also, there will likely be many use cases where manual intervention needs to make sure a task is completed correctly. Finally, there may be processes that add little value to the business, as Guillermo Diaz puts it, "If you start with a pile of garbage, by automating it all you're going to do is make the garbage go faster."

I'm also not advocating violating individual privacy in the name of measuring employee productivity. As Anders Romare reminds us, not only do we need to respect employee privacy, but we'll also likely get pushback from unions if we start monitoring where individuals spend their time too closely.

Of course, privacy is a tricky issue in this social media-infused world where our personal data cascades from platform to platform. For instance, Anders gives the example that Novo Nordisk sells medicine that addresses weight gain that a consumer could discover through a Google search; however, it can be disconcerting for that same consumer (after conducting a Google search regarding the Novo Nordisk product) to log into Facebook and see an ad for a weight control product.

Getting back to the employee privacy concern, one approach is to make the employee data anonymous, which could be done through the ETM application processes.

In any case, the objective should be to reduce Productivity Zone spend through automation and optimization so ET processes leverage data to continuously improve upon themselves. Success would be defined when Productivity Zone ET processes are, for the most part, automated and self-driving. This way, CIOs could devote a vast majority of IT's budget to adding more strategic value by implementing Transformation and Performance Zone work.

The Next CIO

Unfortunately, CIOs have a history of being under funded. The CIO goes to the CFO and asks for $100M to do all the work required to support the business and receives $50M instead. Consequently, the IT organization ends up only managing core support technologies like ERP, CRM, collaboration, email, and networking because that's all they can afford to do given the funding they've been given.

Then, as products have been transitioning from hardware that needs to be shipped to software delivered digitally, the CIOs have been nowhere to be found, partially because they didn't have the budget to participate. This work ended up going to the line of business that started developing really cool products, functionality and innovation.

But then many product groups realized at scale that they needed help managing the availability, performance, security and change management of these products and voila, DevOps started sprouting up in the product groups. This high level of freedom and responsibility to do what they need to do provides the product groups with much more agility. This is especially true when the technology needed to meet their requirements is very specific to the line of business. In essence, each product group leveraging technology on its own has its own CIO equivalent, which is often a businessperson with less knowledge about technology.

This said, there are core, general technologies and services that are more efficiently managed by a centralized organization, like IT. Database is a great example. Generally speaking, it's less efficient for each product group to become an expert in databases, and more efficient for one centralized IT organization to offer a selection of supported databases for the product group to choose from.

On one hand, this is where the CIO and IT organization are headed: Owning the performance, availability, security and change management of technologies and services core to the business, while being subject matter experts helping the lines of business manage the performance, availability, security and change management of their technology choices.

But there's also a sort of stigma that pervades the name IT. Perhaps because so many people never really understood what IT does. Or years of underfunding left IT outside looking in on much of the innovation happening in the business units.

In any case, perhaps as Edward Wustenhoff muses, "What if the next CIO is not called CIO? What if IT is no longer called IT?"

Chief Enterprise Technology Officer and the ET Organization

Afterall, if the CIO is mostly responsible for the core technologies needed to run the business, then perhaps like how the CISO emerged as the Chief Information Security Officer, the CIO's name will transition to something like the Chief Enterprise Technology Officer or CETO?

Likewise, perhaps IT will get rebranded as ET for the Enterprise Technology organization, running a services-based, operating model?

The CETO would manage an ET process catalog populated with all the end-to-end processes the ET organization is responsible for providing to the business, such as a *quote-to-cash* sales process, a *terminated-to-removed* employee offboarding process, and a *demand-to-refresh* asset lifecycle process. These processes would directly align with the solutions the ET organization delivers to the business.

Leveraging an ETM application, the CETO would design and document the workflows required to run these processes, as well as track the relevant KPIs and notifications to drive continuous improvement and articulate value delivered back to the business.

From a line of business perspective, they would see that the ET organization provides a services catalog filled with services the business units need to be successful. These services would all be contained within the ET processes. The ET organization would be responsible for managing the performance, availability, security and change management of the systems and technologies needed to deliver these services.

As an alternative to providing a services catalog, the ET organization could publish a solutions catalog that aligns directly with the delivered ET processes. For instance, instead of publishing a quoting service, the ET organization would provide a *quote-to-cash* solution. Or, of course, the ET organization could publish both.

In this future world, an ETM application would be the key enabler helping the CETO and ET organization not only manage all enterprise technology processes, but also enable the improvement, maturity and optimization of each process along the journey toward running an autonomous ET organization.

Regardless of whether the CIO name changes or not, as Alain Brouhard puts it, an ETM application would enable "the new CIO to be a business transformational leader who excels in leveraging exponential technology at the service of the business."

Challenge to the industry to bring ETM applications to market

In summary, the CIO (and the entire business quite frankly) needs an application that enables them to build and optimize the automated, end-to-end ET processes that touch the ever-expanding enterprise technology portfolio. That application is ETM.

There's clearly a need. The budget wasted on running immature and inefficient ET processes — money wasted that could be redirected to funding disruptive initiatives like digital transformation — is just one of the pain points an ETM application could address. Likewise, by leveraging an ETM application to run a smarter, cost-reducing IT organization, CIOs can also improve productivity, data security and business observability, enabling them to better articulate IT's value creation back to the business.

As Alain Brouhard observes, "An ETM application could power autonomous IT, but not only that, ETM has the power to a) create a new type of "industry solidarity" when it comes to cyber security and b) enable the delivery of real business competitive advantage when it comes to digitization and technology levers."

The time is now for the industry to build ETM applications. Who's in?

Glossary of Terms

Artificial intelligence and machine learning (AI/ML): An "intelligent" computer uses AI to think like a human and perform tasks on its own. Machine learning is how a computer system develops its intelligence. One way to train a computer to mimic human reasoning is to use a neural network, which is a series of algorithms that are modeled after the human brain.

Autonomous IT: An aspiration to describe, automate (to the extent possible and reasonable) and continuously optimize the full range of enterprise technology processes managed by the IT organization.

CISO (Chief Information Security Officer): A senior-level executive responsible for developing and implementing an information security program, which includes procedures and policies designed to protect enterprise communications, systems and assets from both internal and external threats.

CMDB (Configuration management database): an ITIL term for a database used by an organization to store information about hardware and software assets. It is useful to break down configuration items into logical layers.

CRM (Customer Relationship Management) software: Any tool, strategy, or process that helps businesses better organize and access customer data.

Cryptocurrency: a digital currency in which transactions are verified and records maintained by a decentralized system using cryptography, rather than by a centralized authority.

Digital Transformation: The integration of digital technology into all areas of a business, fundamentally changing how a business operates and delivers value to its stakeholders such as customers, partners and employees

Employee offboarding process: The process that leads to the formal separation between an employee and a company through resignation, termination, or retirement. It encompasses all the decisions and processes that take place when an employee leaves.

Employee onboarding process: The processes in which new hires are integrated into the organization. It includes activities that allow new employees to complete an initial new-hire orientation process, as well as learn about the organization and its structure, culture, vision, mission and values.

Enterprise Technology Management Processes consists of five general categories that track the technology lifecycle:
- **Purchase Management** (*plan-to-procure*): The processes required to manage the acquisition of technology to be used by the organization.
- **Deploy Management** (*receive-to-utilize*): The processes that put purchased technology to use within the organization.
- **Monitor Management** (*observe-to-reconcile*): The processes that ensure the organization continues to attain maximum value at minimum cost from its deployed technology.
- **Secure Management** (*detect-to-remediate*): The processes that act on identified security exposures and enforce policies that have been violated.
- **Maintain Management** (*resolve-to-EOL*): The processes that resolve identified cases involving issues with the use of technology.

Enterprise Technology Portfolio consists of five broad categories:

- **Endpoints**: Physical devices connected to the network that run some form of system management software, such as mobile phones, laptops, and point-of-sale terminals.
- **Networking**: Physical and virtual devices that create the network to enable digital communication and interaction between endpoints, servers and storage, such as routers, switches, and firewalls.
- **Infrastructure**: Physical, on-premises servers and storage, as well as cloud-based virtualized machines and storage.
- **Applications**: Software programs installed on endpoints, on-premises or delivered as a service that organizations offer to customers and employees to achieve desired business outcomes.
- **Accessories**: Physical devices that do not connect to the network or run an operating system, such as keyboards and monitors.

ET (Enterprise Technology): The technology assets, such as endpoints, networking, infrastructure, applications and accessories used to run the business.

ETM (Enterprise Technology Management) application: An application that provides *plan-to-EOL* enterprise technology process management, enabling CIOs and the IT organization to define, automate and optimize the end-to-end processes that derive value from the entire often siloed enterprise technology portfolio throughout the technology lifecycle.

ETM Framework: A simple but powerful framework to help focus our attention on the fact that there are processes the company uses to run the business that touch the broad inventory of technology, including:

- **Enterprise Technology Management Processes** are the processes used to run the business that touch the company's enterprise technology portfolio.
- **Enterprise Technology Portfolio** is the entire inventory of technology used by the business.

ET Process Maturity Framework: A five-level framework to asset CIOs and the IT organization:
- Level 1: No process
- Level 2: Described
- Level 3: Partially Automated
- Level 4: Fully Automated
- Level 5: Continuously Automated

Enterprise Technology Maturity: The maturity of the enterprise technology processes deployed by the IT organization and falls into one of five levels:
- **Level 1: No Process** – The process is undefined. Work is done completely manually, is ad hoc and reactive. This is the most inefficient state to be, so Level 1 gets a bad rating.
- **Level 2: Described** – This process is described but implemented manually. Manual processes are inherently inefficient, so Level 2 gets a poor rating, as completely manual processes in general are more expensive and take more time than letting computers do the work through automation.
- **Level 3: Partially Automated** – This process utilizes automation to make it more efficient, but still depends on manual intervention to complete. Because some level of automation is involved, Level 3 gets a fair rating.
- **Level 4: Fully Automated** – This process is completely automated and completion time is measured. Given the process

requires no inefficient manual intervention, Level 4 receives a good rating.

- **Level 5: Continuously Optimized** – The process uses data to optimize and continuously improve its activities. This process is executed, monitored, and managed through dynamic workflows, making decisions based on output from automated tasks, adapting to changing circumstances and conditions, and simultaneously coordinating multiple tasks. Level 5 receives an excellent rating, as processes at this level are running as efficiently and securely as possible.

HRM (Human Resource Management) System: A form of Human Resources (HR) software that combines several systems and processes to ensure the easy management of human resources, business processes and data.

Incumbent organization: An existing organization with some level of market presence

IOT (Internet of Things): Describes the network of physical objects—"things"—that are embedded with sensors, software, and other technologies for the purpose of connecting and exchanging data with other devices and systems over the internet.

ITSM (Information technology service management): The activities that are performed by an organization to design, build, deliver, operate and control information technology services offered to customers.

KPI: Key performance indicator

Multi-cloud environment: A company's use of multiple cloud computing and storage services from different vendors in a single heterogeneous architecture to improve cloud infrastructure capabilities and cost. It also refers to the distribution of cloud assets, software, applications, etc. across several cloud-hosting environments.

POS (Point of Sale) device: a hardware system for processing card payments at retail locations. Software to read magnetic strips of credit and debit cards is embedded in the hardware.

SD-WAN (Software-defined wide area network): A wide area network that uses software-defined network technology, such as communicating over the Internet using overlay tunnels which are encrypted when destined for internal organization locations.

SIEM (Security Information Event Management): Security information and event management (SIEM) technology supports threat detection, compliance and security incident management through the collection and analysis (both near real time and historical) of security events, as well as a wide variety of other event and contextual data sources.

SLA (Service Level Agreement): A commitment between a service provider and a client. Specific aspects of the service – such as quality, availability, responsibilities – are agreed between the service provider and the service user.

Shadow IT: The use of IT-related hardware or software by a department or individual without the knowledge of the IT or security group within the organization. It can encompass cloud services, software, and hardware.

TBM (Technology Business Management): A discipline that improves business outcomes by giving organizations a consistent way to translate technology investments to business value.

Quality of experience (QoE): A measure of the delight or annoyance of a customer's experiences with a service. QoE focuses on the entire service experience; it is a holistic concept, similar to the field of user experience, but with its roots in telecommunication.

Web 3.0: The third generation of the evolution of web technologies. The web, also known as the World Wide Web, is the foundational layer for how the internet is used, providing website and application services.

XLA (Experience Level Agreement): A commitment to delivering a defined experience, ensuring that all service interactions and touchpoints are considered when defining whether the service meets the agreed performance level.

Notes

[i] DJCAPPUCCIO LLC – August 2022

[ii] Think Fast: https://unbounce.com/page-speed-report/

[iii] Number of digital buyers worldwide from 2014 to 2021: https://www.statista.com/statistics/251666/number-of-digital-buyers-worldwide/

[4] Zone to Win website: https://zonetowin.com/

[5] The complexity of more cybersecurity tools: https://www.cyber-observer.com/complexity-of-more-cybersecurity-tools/

[6] 2022 State of ITSM and ESM Survey Report: https://www.informationweek.com/whitepaper/it-strategy/it-leadership/2022-state-of-itsm-and-esm-survey-report/437103

[7] 8 Things Leaders Do That Make Employees Quit: https://hbr.org/2019/09/8-things-leaders-do-that-make-employees-quit

[8] How to calculate cost-per-hire: https://www.glassdoor.com/employers/blog/calculate-cost-per-hire/

[9] The 18 CIS Critical Security Controls: https://www.cisecurity.org/controls/cis-controls-list

[10] Cisco 2020 CISO Benchmark Report: https://www.cisco.com/c/en/us/products/security/ciso-benchmark-report-2020.html

[11] Forbes: 90% Of Companies Have A Multicloud Destiny: Can Conventional Analytics Keep Up? https://www.forbes.com/sites/googlecloud/2022/03/04/90-of-companies-have-a-multicloud-destiny-can-conventional-analytics-keep-up/?sh=20ee5d605d89

[12] CyberSecurity Insiders: Attack Surface Management Maturity Report: https://www.cybersecurity-insiders.com/portfolio/2022-attack-surface-management-maturity-report-oomnitza/

[13] Zone to Win, page 70

[14] 2022 Connectivity Benchmark Report: https://library.mulesoft.com/f276aa86-ed62-4497-ba26-900853024a30

[15] Precision: Principles, Practices and Solutions for the Internet of Things: https://www.amazon.com/Precision-Principles-Practices-Solutions-Internet/dp/1329843568/

[16] NPR: Hackers Disrupt payroll for thousands of employees – including employees: https://www.npr.org/2022/01/15/1072846933/kronos-hack-lawsuits

[17] Cisco Annual Internet Report (2018–2023) White Paper: https://www.cisco.com/c/en/us/solutions/collateral/executive-perspectives/annual-internet-report/white-paper-c11-741490.html

[18] TechTarget: CMDB (Configuration Management Database): https://www.techtarget.com/searchdatacenter/definition/configuration-management-database

[19] Guide to Configuration Management Databases (CMDBs): https://www.atlassian.com/itsm/it-asset-management/cmdb

[20] Gartner Says More Than Half of Enterprise IT Spending in Key Market Segments Will Shift to the Cloud by 2025: https://www.gartner.com/en/newsroom/press-releases/2022-02-09-gartner-says-more-than-half-of-enterprise-it-spending

[21] Gartner Forecasts Worldwide Public Cloud End-User Spending to Reach Nearly $500 Billion in 2022: https://www.gartner.com/en/newsroom/press-releases/2022-04-19-gartner-forecasts-worldwide-public-cloud-end-user-spending-to-reach-nearly-500-billion-in-2022

[22] IT Service Management: https://en.wikipedia.org/wiki/IT_service_management

[23] Ready for ITSM at high velocity: https://www.atlassian.com/itsm/it-asset-management/cmdb

[24] IT Asset Management: the set of business practices that join financial, contractual, and inventory functions to support life cycle management and strategic decision making for the IT environment: https://en.wikipedia.org/wiki/Asset_management

[25] The IT Service Management Survey 2017: https://www.ciowatercooler.co.uk/the-it-service-management-survey-2017/

[26] What is TBM: https://www.tbmcouncil.org/learn-tbm/what-is-tbm/

[27] Gartner Survey Finds Rise in Business Technologists is Driving Funding for Tech Purchases Outside of IT: https://www.gartner.com/en/newsroom/press-releases/2022-03-13-gartner-survey-finds-rise-in-business-technologists-is-driving-funding-for-tech-purchases-outside-of-it

[28] Harvard Business Review Recognizing Employees Is the Simplest Way to Improve Morale: https://hbr.org/2016/05/recognizing-employees-is-the-simplest-way-to-improve-morale.html

[29] 10 Ways to Increase Employee Productivity with Rewards and Recognition: https://www.fond.co/blog/increase-employee-productivity/

Made in the USA
Monee, IL
16 November 2022

54074cbf-08de-412f-8968-cfb93c888c7cR01